SCHAUM'S OUTLINE OF

THEORY AND PROBLEMS

of

ELECTRONIC COMMUNICATION

•

by

LLOYD TEMES, Ph.D., P.E.
Department of Electric Technology
College of Staten Island
City University of New York

SCHAUM'S OUTLINE SERIES
McGRAW-HILL BOOK COMPANY
New York St. Louis San Francisco Auckland Bogotá Düsseldorf Johannesburg London Madrid Mexico Montreal New Delhi Panama Paris São Paulo Singapore Sydney Tokyo Toronto

0-07-063495-5

5 6 7 8 9 10 11 12 13 14 15 SH SH 8 7 6 5

Library of Congress Cataloging in Publication Data

Temes, Lloyd.
Schaum's outline of electronic communication.

(Schaum outline series)
Includes index.
1. Telecommunication. 2. Electronics. I. Title.
II. Title: Electronic communication.
TK5101.T37 621.38 78-11102
ISBN 0-07-063495-5

Preface

This book is intended for use in a first course in communications electronics following a basic course in electronics, which should have included basic device theory. AC circuitry, dc circuitry, and mathematics through intermediate algebra and trigonometry are also necessary prerequisites. Calculus, however, is not used.

This book should prove to be quite complete for a course in communications electronics in a technical training program, and it should be quite helpful in a course for engineers. In addition, it can be used in the preparation for Federal Communications Commission First- and Second-Class Radiotelephone Operations License examinations.

Please note that π has been taken to be 3.14 so as to allow for solutions on calculators not having a special π key.

AM and FM theory and equipment are discussed and dealt with early in the book. There is a chapter on television, and chapters relating to circuits and systems which are usable with the various modulation systems then follow.

I would like to thank Mr. Gilbert Bank of the College of Staten Island for assistance with the text material and with problem solving.

LLOYD TEMES

Contents

Chapter 1

Amplitude Modulation

INTRODUCTION

Modulation is the process of imposing information contained in a lower-frequency electronic signal onto a higher-frequency signal. The higher-frequency signal is called the *carrier* and the lower-frequency signal is called the *modulating signal.* If the information is imposed on the carrier by causing its amplitude to vary in accordance with the modulating signal, the method is called *amplitude modulation.*

The advantage of transmitting the higher-frequency signal is twofold: First, if all radio stations broadcast simultaneously at audio frequencies, they could not be distinguished from one another and only a jumbled mess would be received. Second, it is found that antennas on the order of magnitude of 5 miles to 5000 miles are necessary for *audio* frequency transmissions.

MATHEMATICAL DESCRIPTION

The mathematical description of the unmodulated carrier wave is

$$A \sin 2\pi f_c t$$

where f_c is the carrier frequency and A is the peak value of the unmodulated carrier.

If, for simplicity, a single audio tone is taken as the modulating signal, it can be represented by

$$B \sin 2\pi f_a t$$

where f_a is the frequency of the audio tone and B is the peak value of the modulating signal (see Fig. 1-1).

The modulated wave can be represented mathematically as the product

$$(A + B \sin 2\pi f_a t)(\sin 2\pi f_c t)$$

where f_a is the frequency of the audio modulating signal and f_c is the frequency of the carrier. Factoring, we get

$$A\left(1 + \frac{B}{A} \sin 2\pi f_a t\right)(\sin 2\pi f_c t)$$

In terms of voltage, we have

$$v = V_c\left(1 + \frac{B}{A} \sin 2\pi f_a t\right)(\sin 2\pi f_c t)$$

where V_c is the peak voltage of the unmodulated carrier, represented by A until now.

Making use of the trigonometric identity

$$(\sin X)(\sin Y) = \tfrac{1}{2} \cos (X - Y) - \tfrac{1}{2} \cos (X + Y)$$

the equation describing the amplitude-modulated wave may be written as

$$\boxed{v = V_c \sin 2\pi f_c t + \frac{mV_c}{2} \cos 2\pi(f_c - f_a)t - \frac{mV_c}{2} \cos 2\pi(f_c + f_a)t}$$

where m is called the *modulation factor* and is defined as

$$m = \frac{\text{peak value of modulating signal}}{\text{peak value of unmodulated carrier}}$$

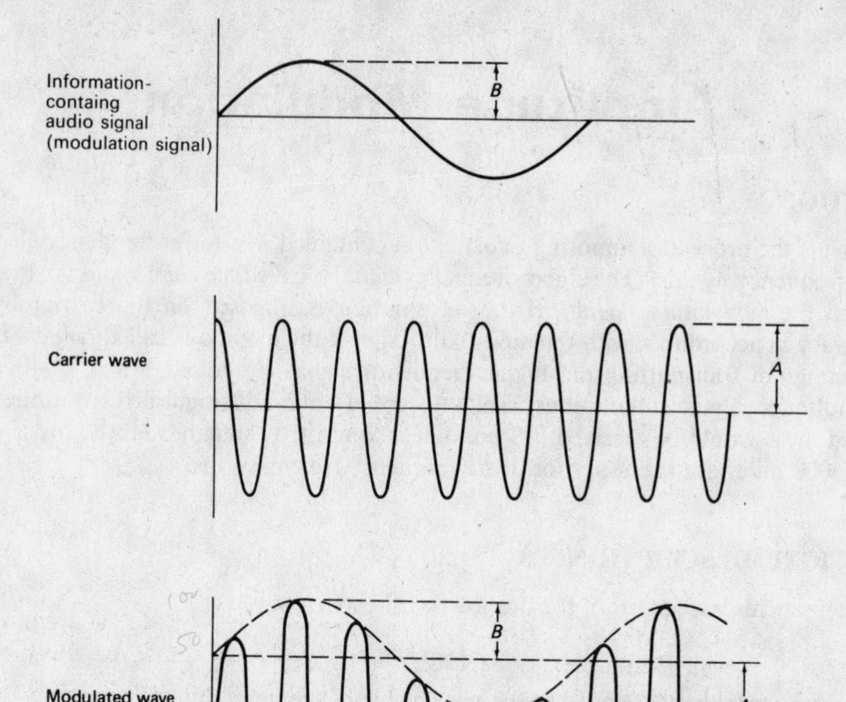

Fig. 1-1

When expressed as a percentage, this is known as the *percent modulation, M*. Using the notation of Fig. 1-1,

$$m = \frac{B}{A}$$

$$M = \frac{B}{A} \times 100\%$$

The percent modulation can vary anywhere from 0 to 100% without introducing distortion. If the percent modulation is allowed to increase beyond 100%, distortion, accompanied by undesirable extraneous frequencies, results. Figure 1-2 depicts three degrees of modulation: (*a*) undermodulation ($M < 100\%$), (*b*) 100% modulation, and (*c*) overmodulation ($M > 100\%$).

Referring to the equation above describing the amplitude-modulated wave, the modulated wave is seen to have three components: one at a frequency of f_c, one at a frequency of $f_c + f_a$, and one at a frequency of $f_c - f_a$ producing the frequency-versus-voltage spectrum as shown in Fig. 1-3(*a*).

The frequency $f_c + f_a$ is called the *upper-side frequency*, and $f_c - f_a$ is called the *lower-side frequency*. Most audio information to be broadcast does not consist of a single pure sine wave. In most cases, a rather complex waveshape is encountered. Any complex waveshape can be considered to be the sum of a set of pure sine waves.

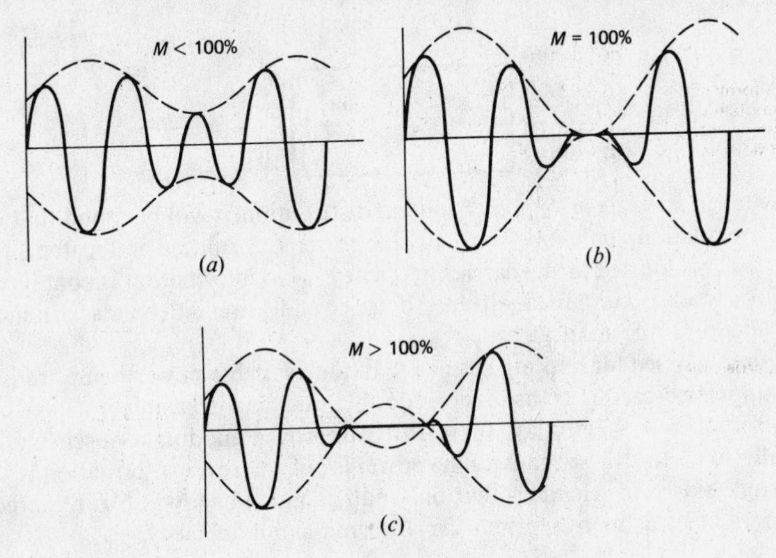

Fig. 1-2

Thus, each of the sine waves which make up the complex audio wave will have both an upper- and a lower-side frequency, which will appear in the analysis of the modulated wave. Rather than discuss an upper- and a lower-side frequency, an upper and a lower sideband of frequencies is referred to.

It can therefore be seen that a broadcast station which intends to broadcast information containing frequencies from 0 to 5000 Hz (5 kHz) needs an upper sideband of 5 kHz and a lower sideband of 5 kHz, for a total bandwidth requirement of 10 kHz. Federal Communications Commission regulations allow a bandwidth of 10 kHz to a station in the AM broadcast band.

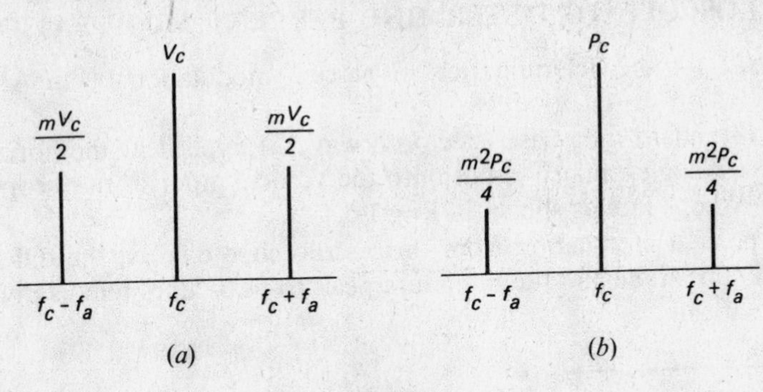

Fig. 1-3

POWER CONTENT

Since power is proportional to the square of voltage, the power-versus-frequency spectrum for an amplitude-modulated wave looks like Fig. 1-3(*b*). Each sideband has a power content equal to $m^2P_c/4$, where P_c is the power content of the signal at the carrier frequency.

Thus the total power is

$$P_T = \frac{m^2P_c}{4} + \frac{m^2P_c}{4} + P_c$$

where P_c is the power content of the carrier and is independent of percent modulation in an AM transmission.

Combining terms,

$$P_T = \frac{m^2P_c}{2} + P_c$$

Factoring,

$$\boxed{P_T = P_c\left(1 + \frac{m^2}{2}\right)}$$

When doing a numerical analysis of power content distribution, it will be found that under optimum conditions (100% modulation) only one-third of the power transmitted is located in the sidebands. Two-thirds of the power is located at the carrier frequency. No information is contained at the carrier frequency. All information is contained within the upper and lower sidebands. In actuality, the two sidebands contain identical information.

Schemes have been devised for making better use of the available power being transmitted. These schemes include suppressed-carrier transmission, double-sideband transmission, and single-sideband transmission. The frequency spectra for these three improved modulation schemes are shown in Fig. 1-4. Essentially, in each of these schemes the power is put where the information is. An additional advantage of the single-sideband scheme is that only half as much bandwidth is required for the transmission, and therefore twice as many stations can transmit simultaneously.

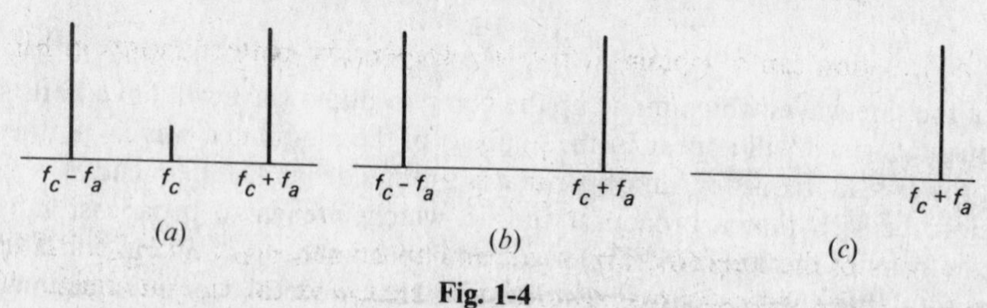

Fig. 1-4

USING THE OSCILLOSCOPE TO DETERMINE PERCENT MODULATION

Two techniques exist for the determination of percent modulation of an AM wave using the oscilloscope.

In one technique, a standard time base (sawtooth wave) is applied to the horizontal input of the 'scope and the AM wave being examined is put onto the vertical input of the 'scope. The AM wave is then displayed on the 'scope face as shown in Fig. 1-5.

To determine the percent modulation from this oscilloscope display, the difference between the maximum peak-to-peak amplitude and the minimum peak-to-peak amplitude is divided by their sum. Examining this mathematically,

$$\boxed{M = \frac{\max p\text{-}p - \min p\text{-}p}{\max p\text{-}p + \min p\text{-}p} \times 100}$$

$$M = \frac{2(A+B) - 2(A-B)}{2(A+B) + 2(A-B)} \times 100$$

$$= \frac{4B}{4A} \times 100$$

$$= \frac{B}{A} \times 100$$

Fig. 1-5

Thus percent modulation can be obtained from the oscilloscope trace shown in Fig. 1-5.

The other technique involves applying the modulated signal to the vertical input of the 'scope and the audio frequency modulating signal to the horizontal input. This results in trapezoidal patterns such as those shown in Fig. 1-6.

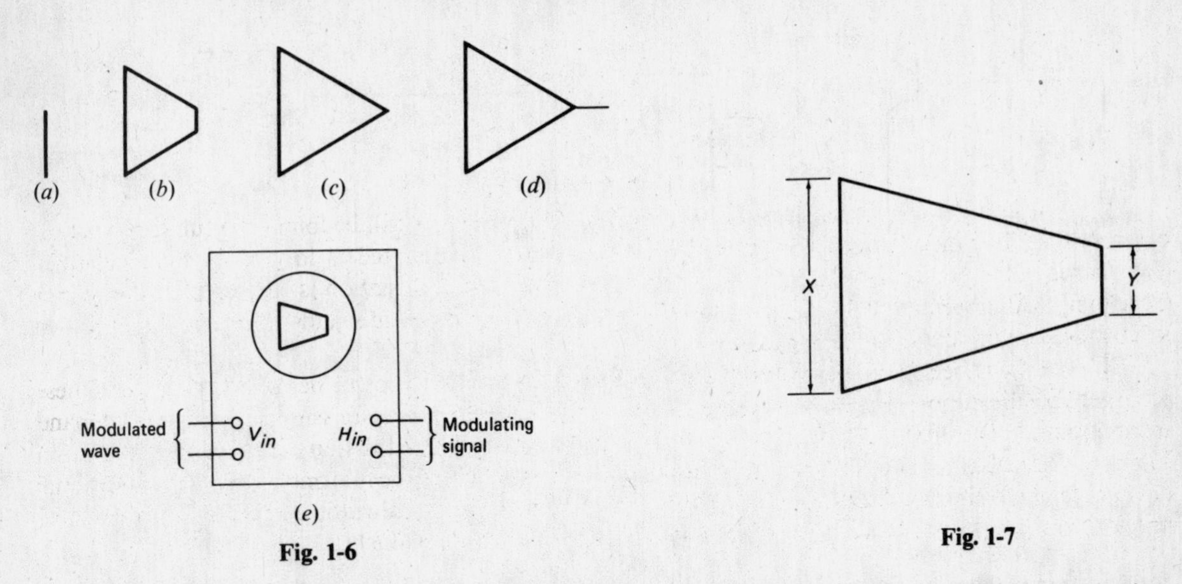

Fig. 1-6

Fig. 1-7

The percent modulation can be obtained from the trapezoidal pattern shown in Fig. 1-7 by using the formula

$$M = \frac{X - Y}{X + Y} \times 100$$

Sometimes the sides of the trapezoid do not appear to be absolutely straight. If the sides of the trapezoid are nonlinear, this indicates that distortion is present in the output signal. Two examples of nonlinear trapezoid sides are shown in Fig. 1-8.

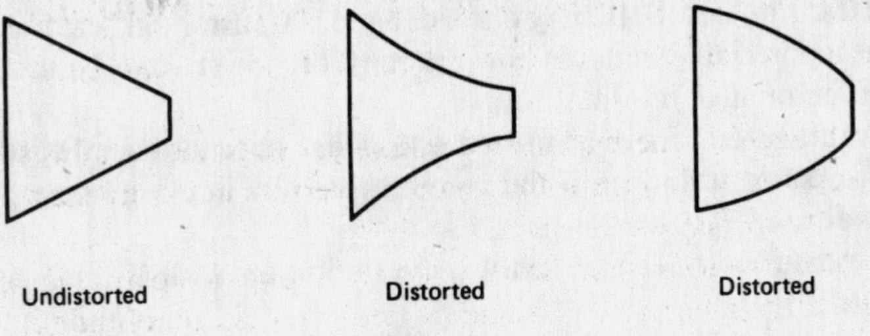

Fig. 1-8

THE AM TRANSMITTER

Figure 1-9 is a block diagram of an AM transmitter. No matter how complicated an AM transmitter may become, it is basically the same as that shown in Fig. 1-9.

It is necessary to have a nonlinear device in the system in order for modulation to occur, that is, to create the sum and difference frequencies necessary for sidebands to appear.

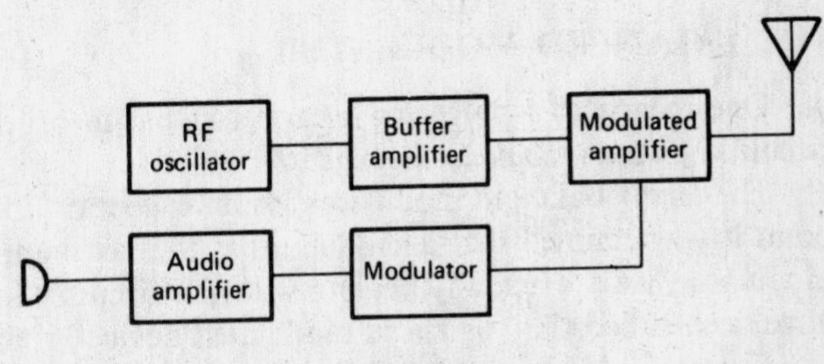

Fig. 1-9

The nonlinear device in which modulation occurs is the *modulated amplifier*.

Figure 1-10 shows two rather simple examples of transistor class C amplifiers which can be used as modulated amplifiers. The fact that these amplifiers are class C means that output current is cut off

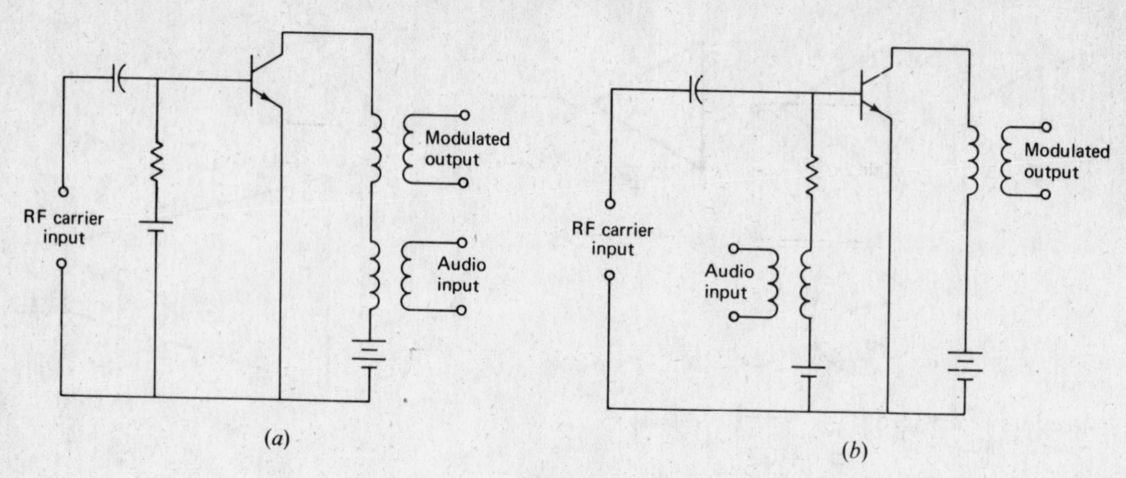

Fig. 1-10

for a portion of the cycle, thereby causing clipping of the output signal. This is then a nonlinear operation and produces the necessary sum and difference frequencies required of an AM transmitter.

Note in particular the terminology used in Fig. 1-9. The modulator is the last stage of audio amplification, while the modulated amplifier is the circuit within which modulation occurs.

SINGLE-SIDEBAND, DOUBLE-SIDEBAND, AND PILOT-CARRIER TRANSMISSION

In single-sideband transmission, SSB, only one sideband is transmitted, leaving off the other sideband and the carrier. This appreciably reduces the amount of power and bandwidth necessary to be transmitted for a given amount of information.

Another added advantage to SSB over standard AM is that since the signal has a narrower bandwidth a narrower passband is permissible within the receiver, thereby limiting noise pickup because of the narrower open bandwidth.

Double-sideband transmission, DSB, is another variation on AM. In this case, only the carrier is eliminated.

Another variation is called pilot-carrier transmission, in which the two sidebands as well as a trace of the carrier are transmitted.

THE BALANCED MODULATOR

One means of suppressing a carrier signal in order to create an SSB or DSB signal is to use a circuit known as a *balanced modulator*.

The term "balanced modulator" is an exception to the rule stated earlier in this chapter. At that point it was claimed that a modulator is the last audio amplifier stage. In the case of the balanced modulator, we are actually dealing with the circuitry in which modulation is taking place. It should have been called the "balanced modulated device" if the original rule were followed.

Figure 1-11 shows some variations of balanced modulators.

Although there are many advantages to SSB, DSB, and pilot-carrier transmission, they have not gained general acceptance for use in home-entertainment equipment but are reserved for point-to-point communications because of the complexity of the equipment and thus the associated increased cost. The reason for the extra cost is the need to reconstitute an AM signal prior to demodulation.

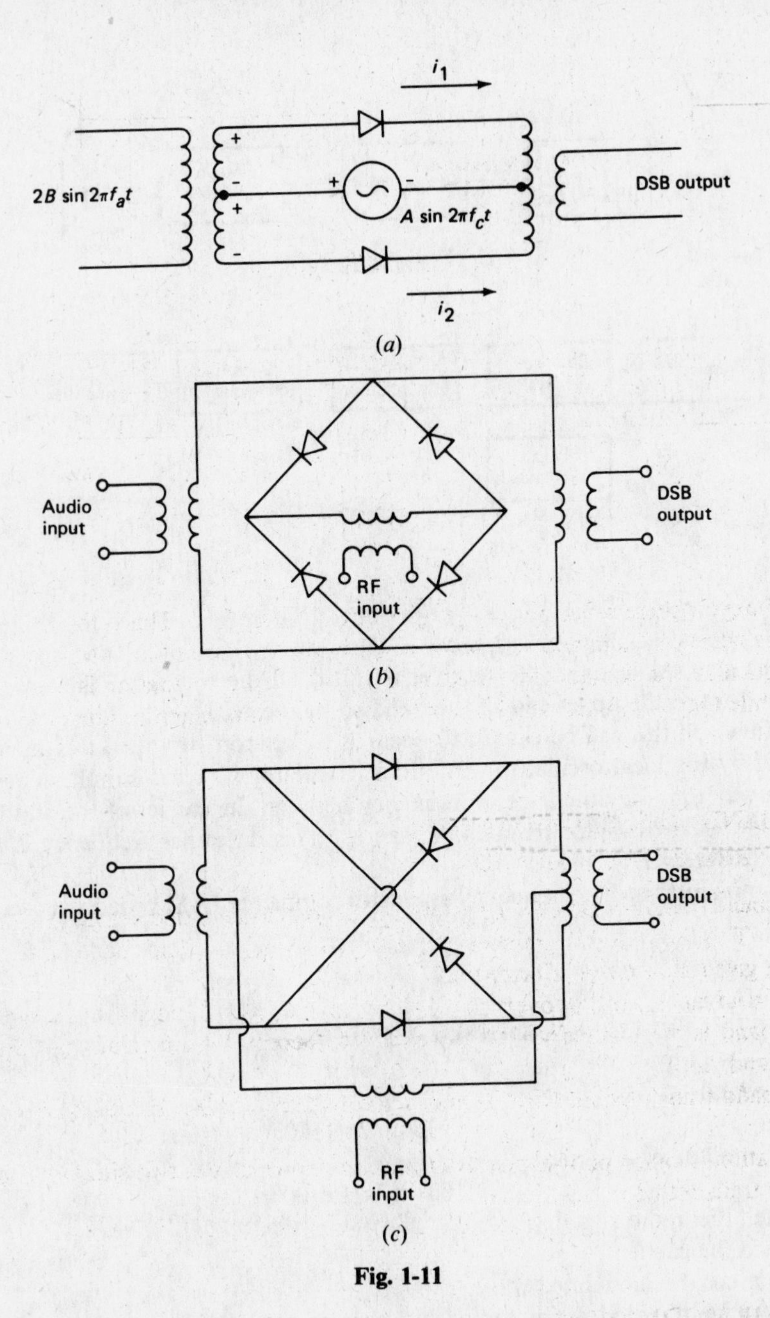

Fig. 1-11

AM RECEIVERS

The simplest AM receiver consists of a tandem arrangement of a selector–RF amplifier combination, a demodulator, an audio amplifier, and a speaker as shown in Fig. 1-12. This simple receiver is called a tuned radio-frequency (TRF) receiver.

The TRF receiver has been replaced for the most part by the superheterodyne receiver except in cases of the least expensive toylike equipment.

The main difficulty with the TRF receiver is the varying passband encountered as the receiver is tuned from the low end of the frequency band to be received to the high end of the frequency band to be received.

In the superheterodyne receiver, a block diagram of which is shown as Fig. 1-13, most of the high-frequency amplification takes place in the intermediate-frequency section, the passband and center

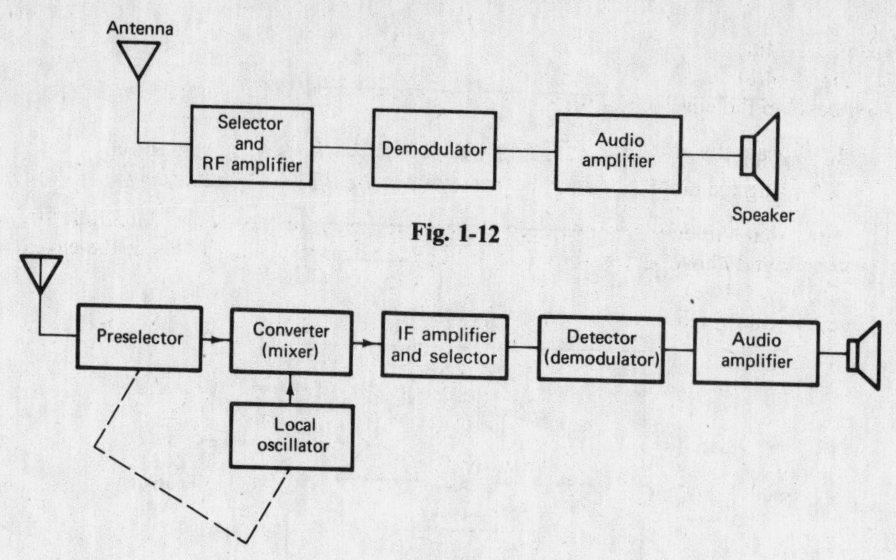

Fig. 1-12

Fig. 1-13

frequency of which stay the same as the receiver is tuned all the way from the lower portion of the band to be received up to the upper end of the band to be received.

The manner in which this can be caused to occur is to heterodyne (beat) the incoming signal with a signal generated by the local oscillator. The local oscillator is tuned simultaneously with the RF selector in such a way that the difference in frequency between the carrier of the station tuned to and the frequency of the local oscillator remains the same. This difference frequency is the intermediate frequency, IF, of the receiver.

The most common intermediate frequency used with commercial AM broadcast receivers is 455 kHz.

Solved Problems

1.1 An audio signal

$$15 \sin 2\pi(1500t)$$

amplitude modulates a carrier

$$60 \sin 2\pi(100\,000t)$$

(*a*) Sketch the audio signal.
(*b*) Sketch the carrier.
(*c*) Construct the modulated wave.
(*d*) Determine the modulation factor and percent modulation.
(*e*) What are the frequencies of the audio signal and the carrier?
(*f*) What frequencies would show up in a spectrum analysis of the modulated wave?

SOLUTION

Given: Audio signal $= 15 \sin 2\pi(1500t)$
Carrier $= 60 \sin 2\pi(100\,000t)$

Find: (*a*) Sketch of audio signal
(*b*) Sketch of carrier
(*c*) Sketch of modulated wave
(*d*) m, M
(*e*) f_a, f_c
(*f*) Frequency content of modulated wave

(*a*) See Fig. 1-14(*a*).
(*b*) See Fig. 1-14(*b*).
(*c*) First develop the envelope of the modulated wave:

1. Locate the amplitude of the carrier (dashed line).
2. Using the amplitude of the carrier as an axis, lay in the audio signal.

Now that the envelope has been determined, a signal having an amplitude defined by the envelope found above and having a frequency of the carrier is laid in within the envelope. See Fig. 1-14(*c*).

(*d*) Using the following equation for modulation factor,

$$m = \frac{\text{audio amplitude}}{\text{carrier amplitude}} = \frac{B}{A}$$
$$= \tfrac{15}{60} = \tfrac{1}{4}$$

$$\boxed{m = 0.25}$$

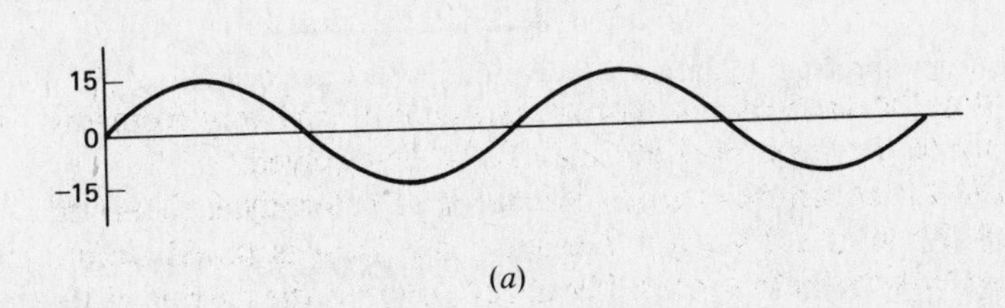

(*a*)

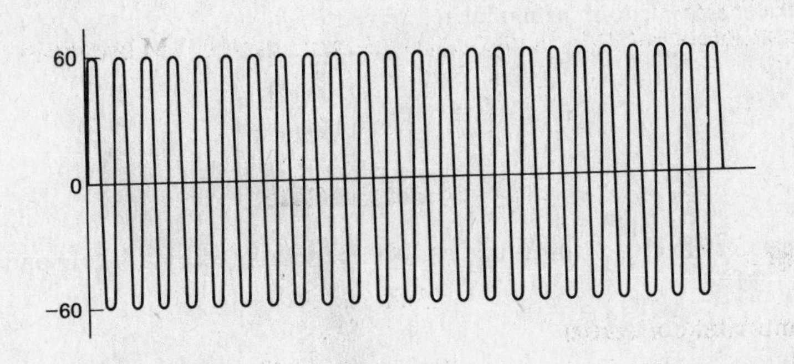

(*b*)

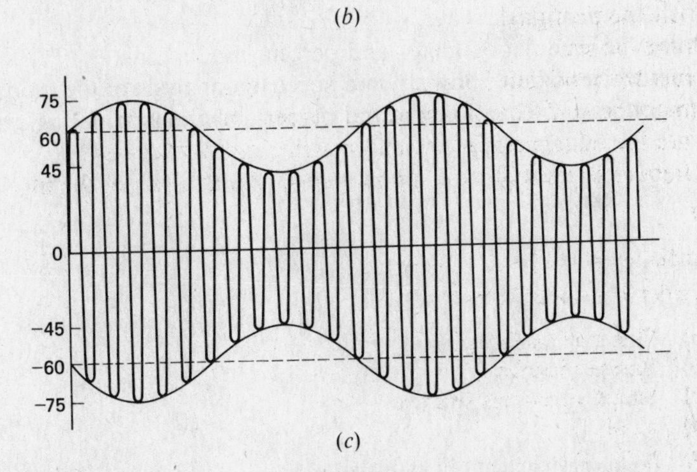

(*c*)

Fig. 1-14

Converting modulation factor to percent modulation,

$$M = m \times 100$$
$$= 0.25 \times 100$$

$$\boxed{M = 25\%}$$

(*e*) Since

$$\text{Audio signal} = B \sin 2\pi f_a t$$
$$= 15 \sin 2\pi(1500t)$$

$$\boxed{f_a = 1500 \text{ Hz}}$$

Since

$$\text{Carrier} = A \sin 2\pi f_c t$$
$$= 60 \sin 2\pi(100\,000t)$$

$$\boxed{f_c = 100\,000 \text{ Hz}}$$

(*f*) The frequency spectrum of an amplitude-modulated wave consists of

$$f_c, \qquad f_c + f_a, \qquad \text{and} \qquad f_c - f_a$$

$$f_c = 100\,000 \text{ Hz}$$

$$f_c + f_a = 100\,000 + 1500 = 101\,500 \text{ Hz}$$

$$f_c - f_a = 100\,000 - 1500 = 98\,500 \text{ Hz}$$

The frequency content of the modulated wave is

$$\boxed{\begin{array}{c} 100\,000 \text{ Hz} \\ 101\,500 \text{ Hz} \\ 98\,500 \text{ Hz} \end{array}}$$

1.2 A 75-MHz carrier having an amplitude of 50 V is modulated by a 3-kHz audio signal having an amplitude of 20 V.

(*a*) Sketch the audio signal.
(*b*) Sketch the carrier.
(*c*) Construct the modulated wave.
(*d*) Determine the modulation factor and percent modulation.
(*e*) What frequencies would show up in a spectrum analysis of the modulated wave?
(*f*) Write trigonometric equations for the carrier and the modulating waves.

SOLUTION

Given:
$f_c = 75$ MHz
$A = 50$ V
$f_a = 3$ kHz
$B = 20$ V

Find:
(*a*) Sketch of audio signal
(*b*) Sketch of carrier
(*c*) Sketch of modulated wave
(*d*) m, M
(*e*) Frequency content of modulated wave
(*f*) Trigonometric equations for modulating signal and carrier

(*a*) See Fig. 1-15(*a*).

(*b*) See Fig. 1-15(*b*).

(*c*) The envelope of the carrier is first developed by drawing the horizontal dashed line at the unmodulated carrier amplitude, both positive and negative. The audio signal is now sketched around the dashed line, providing the envelope within which the radio-frequency signal can be laid in. See Fig. 1-15(*c*).

(*d*) From the defining equation of modulation factor,

$$m = \frac{B}{A} = \frac{20}{50}$$

$$\boxed{m = 0.4}$$

Percent modulation can now be determined by multiplying the modulation factor by 100:

$$M = m \times 100 = 0.4 \times 100$$

$$\boxed{M = 40\%}$$

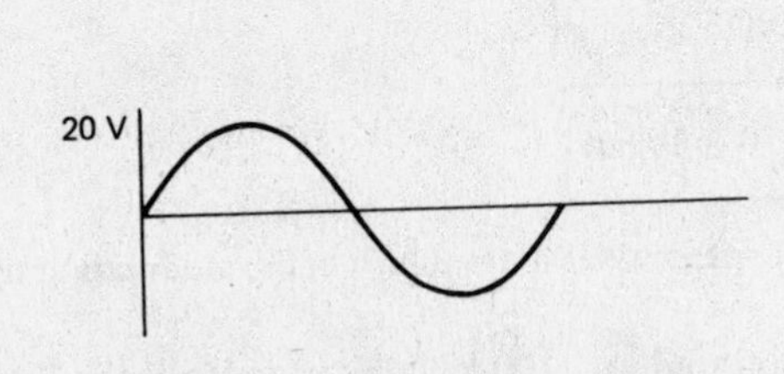

(*a*)

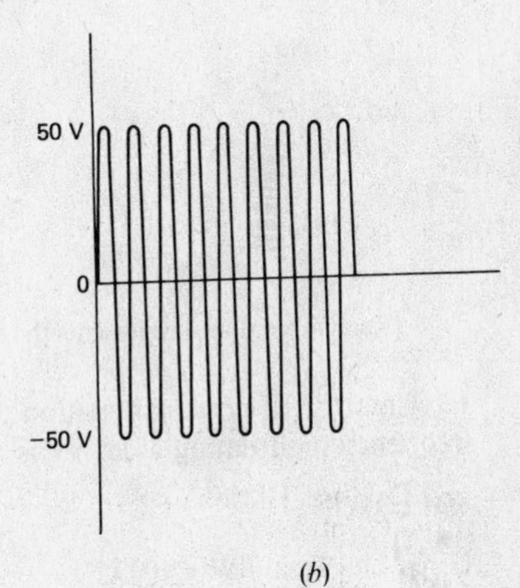

(*b*)

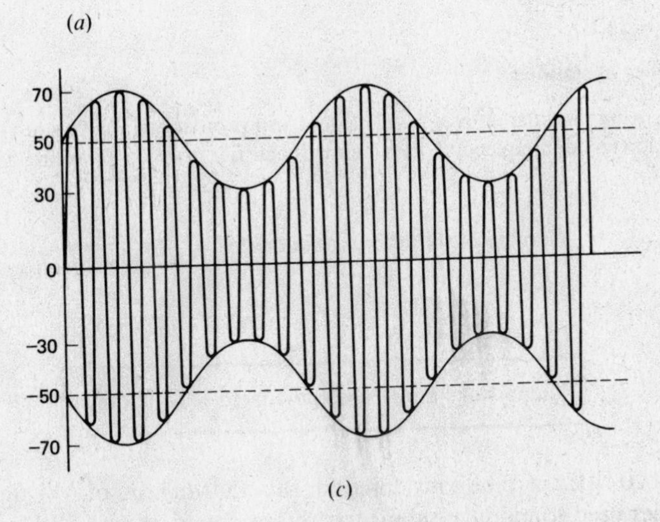

(*c*)

Fig. 1-15

(*e*) The frequency content of an AM signal consists of the carrier frequency and the side frequencies which result from adding the audio frequency to the carrier and from subtracting the audio frequency from the carrier frequency.

$$f_c = 75\text{ MHz}$$

$$\begin{aligned} f_c + f_a &= 75\text{ MHz} + 3\text{ kHz} \\ &= 75\,000\text{ kHz} + 3\text{ kHz} \\ &= 75\,003\text{ kHz} \end{aligned}$$

$$\begin{aligned} f_c - f_a &= 75\,000\text{ kHz} - 3\text{ kHz} \\ &= 74\,997\text{ kHz} \end{aligned}$$

Thus, the frequency content of the AM wave is

$$\boxed{\begin{array}{c} 75.000\text{ MHz} \\ 75.003\text{ MHz} \\ 74.997\text{ MHz} \end{array}}$$

(*f*)

$$\begin{aligned} v_a &= B \sin 2\pi f_a t \\ &= 20 \sin 2\pi(3000)t \end{aligned}$$

$$\boxed{v_a = 20 \sin 6000\pi t}$$

$$\begin{aligned} v_c &= A \sin 2\pi f_c t \\ &= 50 \sin 2\pi(75 \times 10^6)t \end{aligned}$$

$$\boxed{v_c = 50 \sin 150 \times 10^6 \pi t}$$

(Remember that *A* represents the amplitude of the carrier and *B* the amplitude of the modulating signal.)

1.3 How many AM broadcast stations can be accommodated in a 100-kHz bandwidth if the highest frequency modulating a carrier is 5 kHz?

SOLUTION

Given: Total BW = 100 kHz
$f_{a\,\max} = 5$ kHz

Find: Number of stations

Any station being modulated by a 5-kHz signal will produce an upper-side frequency 5 kHz above its carrier and a lower-side frequency 5 kHz below its carrier, thereby requiring a bandwidth of 10 kHz. Thus,

$$\begin{aligned} \text{Number of stations accommodated} &= \frac{\text{total BW}}{\text{BW per station}} \\ &= \frac{100 \times 10^3}{10 \times 10^3} \end{aligned}$$

$$\boxed{\text{Number of stations accommodated} = 10\text{ stations}}$$

1.4 A bandwidth of 20 MHz is to be considered for the transmission of AM signals. If the highest audio frequencies used to modulate the carriers are not to exceed 3 kHz, how many stations could broadcast within this band simultaneously without interfering with one another?

SOLUTION

Given: Total BW = 20 MHz

$f_{a\,max} = 3$ kHz

Find: Number of AM stations

The maximum bandwidth of each AM station is determined by the maximum frequency of the modulating signal.

$$\text{Station BW} = 2f_{a\,max} = 2 \times 3 \times 10^3 = 6 \times 10^3 = 6 \text{ kHz}$$

Thus, the number of stations that can broadcast simultaneously without interfering with one another is

$$\frac{20 \times 10^6}{6 \times 10^3} = 3.333 \times 10^3$$

Number of stations = 3333

1.5 The total power content of an AM signal is 1000 W. Determine the power being transmitted at the carrier frequency and at each of the sidebands when the percent modulation is 100%.

SOLUTION

Given: $P_T = 1000$ W

$M = 100\%$; therefore, $m = 1$

Find: P_c, P_{USB}, P_{LSB}

The total power consists of the power at the carrier frequency, that at the upper sideband, and that at the lower sideband.

$$P_T = P_c + P_{USB} + P_{LSB}$$

See Fig. 1-16.

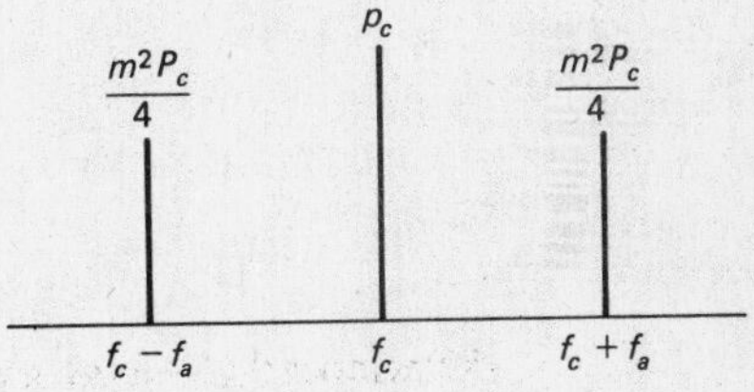

Fig. 1-16

From the equation for total power,

$$P_T = P_c + \frac{m^2P_c}{4} + \frac{m^2P_c}{4}$$

$$= P_c + \frac{m^2P_c}{2}$$

$$1000 = P_c + \frac{(1.0)^2P_c}{2}$$

$$= P_c + 0.5P_c$$

$$= 1.5P_c$$

$$\frac{1000}{1.5} = P_c$$

$$P_c = 666.67 \text{ W}$$

This leaves $1000 - 666.67 = 333.33$ W to be shared equally between upper and lower sidebands.

$$P_{USB} + P_{LSB} = 333.33 \text{ W}$$

$$P_{USB} = P_{LSB}$$

$$2P_{LSB} = 333.33$$

$$P_{LSB} = P_{USB} = \frac{333.33}{2}$$

$$= 166.66$$

$$P_c = 666.67 \text{ W}$$
$$P_{USB} = P_{LSB} = 166.66 \text{ W}$$

1.6 Determine the power content of the carrier and each of the sidebands for an AM signal having a percent modulation of 80% and a total power of 2500 W.

SOLUTION

Given: $M = 80\%$; $m = 0.8$
$P_T = 2500$ W

Find: P_c, P_{USB}, P_{LSB}

The total power of an AM signal is the sum of the power at the carrier frequency and the power contained in the sidebands.

$$P_T = P_c + P_{USB} + P_{LSB}$$

Using the equation for total power,

$$P_T = P_c + \frac{m^2P_c}{4} + \frac{m^2P_c}{4}$$

$$\frac{m^2P_c}{4} + \frac{m^2P_c}{4} = \frac{m^2P_c}{2}$$

$$P_T = P_c + \frac{m^2P_c}{2}$$

$$2500 = P_c + \frac{(0.8)^2P_c}{2}$$

$$= P_c + \frac{0.64}{2}P_c = 1.32P_c$$

$$= 1.32P_c$$

$$P_c = \frac{2500}{1.32}$$

$$P_c = 1893.9 \text{ W}$$

The power in the two sidebands is the difference between the total power and the carrier power.

$$P_{USB} + P_{LSB} = 2500 - 1893.9$$
$$P_{USB} + P_{LSB} = 606.1 \text{ W}$$
$$P_{USB} = P_{LSB} = \frac{606.1}{2} \text{ W}$$

$$P_{USB} = P_{LSB} = 303.50 \text{ W}$$

1.7 The power content of the carrier of an AM wave is 5 kilowatts (kW). Determine the power content of each of the sidebands and the total power transmitted when the carrier is modulated 75%.

SOLUTION

Given: $P_c = 5$ kW
$M = 75\%$; $m = 0.75$

Find: P_{USB}, P_{LSB}, P_T

Since, in an AM wave, the power in each of the sidebands is equal,

$$P_{USB} = P_{LSB} = \frac{m^2 P_c}{4}$$

$$= \frac{(0.75)^2(5000)}{4}$$

$$\boxed{P_{USB} = P_{LSB} = 703.13 \text{ W}}$$

The total power is the sum of the carrier power and the power in the two sidebands.

$$P_T = P_c + P_{USB} + P_{LSB}$$

$$= 5000 + 703.13 + 703.13$$

$$\boxed{P_T = 6406.26 \text{ W}}$$

1.8 An amplitude-modulated wave has a power content of 800 W at its carrier frequency. Determine the power content of each of the sidebands for a 90% modulation.

SOLUTION

Given: $P_c = 800$ W
$M = 90\%$; $m = 0.90$

Find: P_{LSB}, P_{USB}

The power in each of the sidebands is equal to $m^2 P_c/4$.

$$P_{LSB} = P_{USB} = \frac{m^2 P_c}{4}$$

$$= \frac{(0.9)^2 800}{4}$$

$$\boxed{P_{LSB} = P_{USB} = 162 \text{ W}}$$

1.9 Determine the percent modulation of an amplitude-modulated wave which has a power content at the carrier of 8 kW and 2 kW in each of its sidebands when the carrier is modulated by a simple audio tone.

SOLUTION

Given: $P_c = 8$ kW
$P_{USB} = P_{LSB} = 2$ kW

Find: M

Knowing the power content of the sidebands and the carrier, the relationship of sideband power can be used to determine the modulation factor. Once the modulation factor is known, merely multiplying it by 100 provides percent modulation.

$$P_{USB} = P_{LSB} = \frac{m^2 P_c}{4}$$

$$2 \times 10^3 = \frac{m^2(8 \times 10^3)}{4}$$

$$m^2 = \frac{4 \times 2 \times 10^3}{8 \times 10^3}$$
$$= 1.0$$
$$m = 1.0$$
$$M = m \times 100$$
$$\boxed{M = 100\%}$$

1.10 The total power content of an AM wave is 600 W. Determine the percent modulation of the signal if each of the sidebands contains 75 W.

SOLUTION

Given: $P_T = 600$ W

$P_{USB} = P_{LSB} = 75$ W

Find: M

In order to determine the percent modulation, the power contained at the carrier frequency is first determined. Once P_c is known, the relationship between P_c and the sideband power will provide a means of determining the modulation factor, from which the percent modulation is easily found.

Carrier power can be determined from the following:

$$P_T = P_c + P_{USB} + P_{LSB}$$
$$600 = P_c + 75 + 75$$
$$P_c = 600 - 150$$
$$= 450$$

Now using the relationship between sideband power and carrier power,

$$P_{USB} = P_{LSB} = \frac{m^2 P_c}{4}$$
$$75 = \frac{m^2(450)}{4}$$
$$m^2 = \frac{4(75)}{450}$$
$$= 0.667$$
$$m = 0.816$$

Converting modulation factor to percent modulation,

$$M = m \times 100$$
$$= 0.816 \times 100$$
$$\boxed{M = 81.6\%}$$

1.11 Find the percent modulation of an AM wave whose total power content is 2500 W and whose sidebands each contain 400 W.

SOLUTION

Given: $P_T = 2500$ W

$P_{USB} = P_{LSB} = 400$ W

Find: M

First find the power contained at the carrier frequency. Then use the relationship between sideband power and carrier power to determine the modulation factor. Once the modulation factor is known, the percent modulation can easily be found merely by multiplying by 100.

The power at the carrier frequency can be determined from the following:

$$P_T = P_c + P_{\text{USB}} + P_{\text{LSB}}$$
$$2500 = P_c + 400 + 400$$
$$P_c = 2500 - 800$$
$$= 1700 \text{ W}$$
$$P_{\text{USB}} = P_{\text{LSB}} = \frac{m^2 P_c}{4}$$
$$400 = \frac{m^2(1700)}{4}$$
$$m^2 = \frac{400(4)}{1700}$$
$$= \frac{1600}{1700}$$
$$= 0.941$$
$$m = 0.970$$
$$M = 0.970 \times 100$$
$$\boxed{M = 97\%}$$

1.12 Determine the power content of each of the sidebands and of the carrier of an AM signal that has a percent modulation of 85% and contains 1200 W total power.

SOLUTION

Given: $M = 85\%$; $m = 0.85$
$P_T = 1200$ W

Find: P_c, P_{USB}, P_{LSB}

Using the equation which relates total power to carrier power,

$$P_T = P_c\left(1 + \frac{m^2}{2}\right)$$
$$1200 = P_c\left[1 + \frac{(0.85)^2}{2}\right]$$
$$= P_c\left[1 + \frac{0.7225}{2}\right]$$
$$= P_c[1 + 0.3613]$$
$$= 1.3613P_c$$
$$P_c = \frac{1200}{1.3613}$$
$$\boxed{P_c = 881.5 \text{ W}}$$

The sum of carrier power and sideband power is equal to total power.

$$\begin{aligned} P_c + P_{SB} &= P_T \\ 881.5 + P_{SB} &= 1200 \\ P_{SB} &= 1200 - 881.5 \\ &= 318.5 \end{aligned}$$

The total sideband power is made up equally of upper sideband power and lower sideband power.

$$\begin{aligned} P_{USB} = P_{LSB} &= \frac{P_{SB}}{2} \\ &= \frac{318.5}{2} \\ &= 159.25 \end{aligned}$$

$$\boxed{\begin{aligned} P_c &= 881.5 \text{ W} \\ P_{USB} = P_{LSB} &= 159.25 \text{ W} \end{aligned}}$$

1.13 An AM signal in which the carrier is modulated 70% contains 1500 W at the carrier frequency. Determine the power content of the upper and lower sidebands for this percent modulation. Calculate the power at the carrier and the power content of each of the sidebands when the percent modulation drops to 50%.

SOLUTION

Given: $M = 70\%$; $m = 0.70$

$P_{c70} = 1500$ W

Find: $P_{USB_{70}}$, $P_{LSB_{70}}$, P_{c50}, $P_{USB_{50}}$, $P_{LSB_{50}}$

The power content of each of the sidebands is equal to $m^2P_c/4$.

$$\begin{aligned} P_{USB_{70}} = P_{LSB_{70}} &= \frac{m^2P_c}{4} \\ &= \frac{(0.7)^2 1500}{4} \\ &= 183.75 \end{aligned}$$

$$\boxed{P_{USB_{70}} = P_{LSB_{70}} = 183.75 \text{ W}}$$

In standard AM transmission, carrier power remains the same, regardless of percent modulation. Thus,

$$\begin{aligned} P_{c50} &= P_{c70} = 1500 \text{ W} \\ P_{USB_{50}} = P_{LSB_{50}} &= \frac{m^2P_c}{4} \\ &= \frac{(0.5)^2 1500}{4} \\ &= 93.75 \text{ W} \end{aligned}$$

$$\boxed{\begin{aligned} P_{c50} &= 1500 \text{ W} \\ P_{USB_{50}} = P_{LSB_{50}} &= 93.75 \text{ W} \end{aligned}}$$

1.14 The percent modulation of an AM wave changes from 40% to 60%. Originally, the power content at the carrier frequency was 900 W. Determine the power content at the carrier frequency and within each of the sidebands after the percent modulation has risen to 60%.

SOLUTION

Given: $M_1 = 40\%$; $m_1 = 0.40$
$M_2 = 60\%$; $m_2 = 0.60$
$P_{c40} = 900$ W

Find: P_{c60}, P_{USB60}, P_{LSB60}

The power content of the carrier of an AM signal remains the same regardless of percent modulation. Thus,

$$P_{c60} = P_{c40} = 900 \text{ W}$$

The power content of each of the sidebands is equal to $m^2P_c/4$.

$$P_{USB60} = P_{LSB60} = \frac{m^2P_c}{4}$$

$$= \frac{(0.60)^2(900)}{4}$$

$$\boxed{P_{USB60} = P_{LSB60} = 81.0 \text{ W}}$$

1.15 A single-sideband (SSB) signal contains 1 kW. How much power is contained in the sidebands and how much at the carrier frequency?

SOLUTION

Given: $P_{SSB} = 1$ kW

Find: P_{SB}, P_c

In a single-sideband transmission, the carrier and one of the sidebands have been eliminated. Therefore, all the transmitted power is transmitted at one of the sidebands regardless of percent modulation. Thus,

$$\boxed{\begin{aligned} P_{SB} &= 1 \text{ kW} \\ P_c &= 0 \text{ W} \end{aligned}}$$

1.16 An SSB transmission contains 10 kW. This transmission is to be replaced by a standard amplitude-modulated signal with the same power content. Determine the power content of the carrier and each of the sidebands when the percent modulation is 80%.

SOLUTION

Given: $P_{SSB} = 10$ kW
$M = 80\%$; $m = 0.80$

Find: P_c, P_{LSB}, P_{USB}

Since the total power content of the new AM signal is to be the same as the total power content of the SSB signal,

$$P_T = P_{SSB} = 10 \text{ kW}$$

Solving for power contained at the carrier frequency,

$$
\begin{aligned}
P_T &= P_c + P_{LSB} + P_{USB} \\
&= P_c + \frac{m^2 P_c}{4} + \frac{m^2 P_c}{4} \\
10\,000 &= P_c + \frac{(0.8)^2 P_c}{4} + \frac{(0.8)^2 P_c}{4} \\
&= P_c + \frac{0.64 P_c}{2} \\
&= 1.32 P_c \\
\frac{10\,000}{1.32} &= P_c \\
P_c &= 7575.76 \text{ W}
\end{aligned}
$$

The power content of the sidebands is equal to the difference between the total power and the carrier power.

$$P_{SB} = P_T - P_c$$

The power content of the upper and the lower sidebands is equal.

$$
\begin{aligned}
P_{LSB} + P_{USB} &= 10\,000 - 7575.76 \\
&= 2424.24 \\
P_{LSB} = P_{USB} &= \frac{2424.24}{2} \\
&= 1212.12 \text{ W}
\end{aligned}
$$

Thus,

$$
\boxed{\begin{aligned}
P_c &= 7575.76 \text{ W} \\
P_{LSB} = P_{USB} &= 1212.12 \text{ W}
\end{aligned}}
$$

1.17 Determine the modulation factor and percent modulation of the signal shown as Fig. 1-17.

SOLUTION

Given: AM signal as shown in Fig. 1-17

Find: m and M

Using the equation relating maximum peak-to-peak amplitude and minimum peak-to-peak amplitude to modulation factor

$$m = \frac{\max p\text{-}p - \min p\text{-}p}{\max p\text{-}p + \min p\text{-}p}$$

we get from Fig. 1-17

$$\max p\text{-}p = 2(80) = 160$$
$$\min p\text{-}p = 2(20) = 40$$

$$m = \frac{160 - 40}{160 + 40} = \frac{120}{200} = 0.6$$

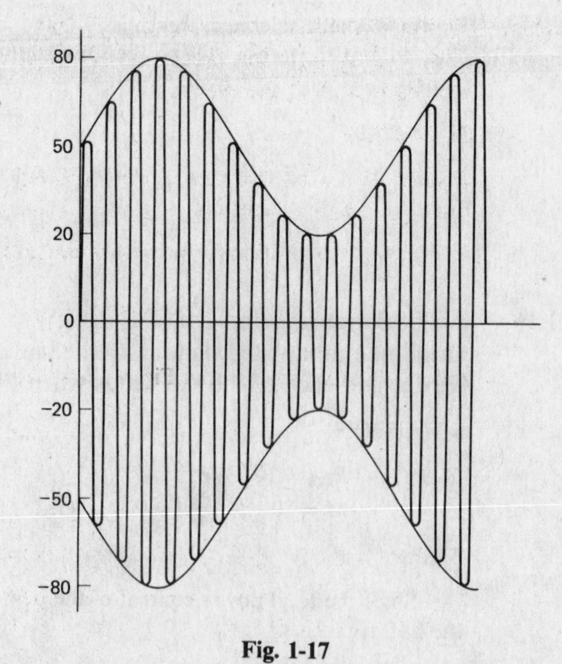

Fig. 1-17

$$\boxed{m = 0.6}$$

$$M = m \times 100 = 0.6 \times 100$$

$$\boxed{M = 60\%}$$

1.18 Find the modulation index and percent modulation of the signal shown as Fig. 1-18.

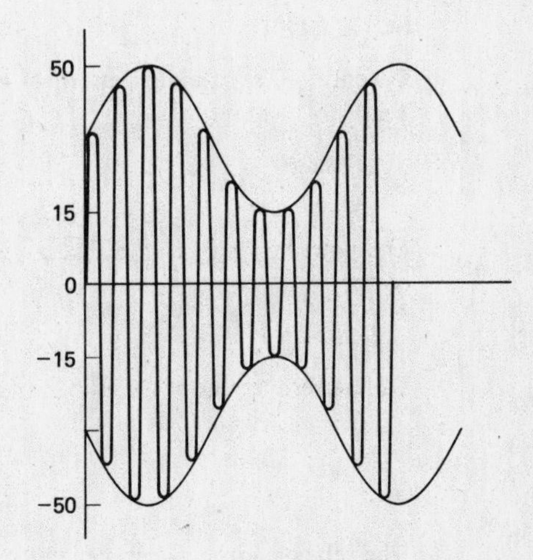

Fig. 1-18

SOLUTION

Given: AM signal as shown in Fig. 1-18
Find: m and M

Using

$$m = \frac{\max p\text{-}p - \min p\text{-}p}{\max p\text{-}p + \min p\text{-}p}$$

and values from Fig. 1-18, we get

$$\max p\text{-}p = 2(50) = 100$$
$$\min p\text{-}p = 2(15) = 30$$

$$m = \frac{100 - 30}{100 + 30} = \frac{70}{130}$$

$$\boxed{m = 0.538}$$

$$M = m \times 100$$

$$\boxed{M = 53.8\%}$$

1.19 The trapezoidal pattern shown in Fig. 1-19 results when examining an AM wave. Determine the modulation factor and percent modulation of the wave. What can be said about the distortion of the AM wave?

SOLUTION

Given: Trapezoidal pattern shown as Fig. 1-19
Find: m, M, distortion

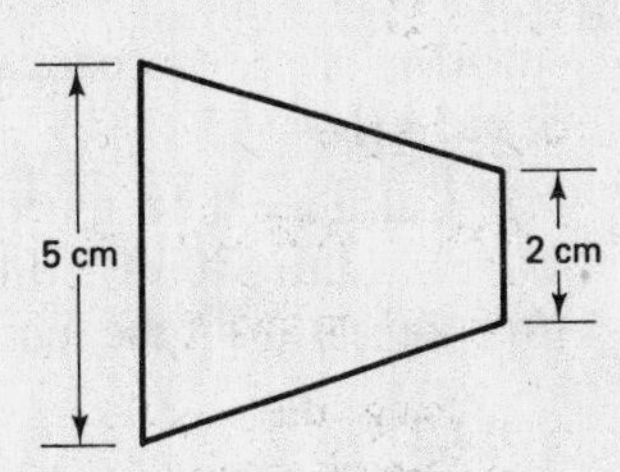

Fig. 1-19

Using the equation for percent modulation,

$$M = \frac{x - y}{x + y} \times 100$$

and substituting values from Fig. 1-19, we get

$$M = \frac{5 - 2}{5 + 2} \times 100$$

$$= \frac{3}{7} \times 100$$

$$\boxed{\begin{aligned} M &= 42.9\% \\ m &= 0.429 \end{aligned}}$$

Regarding distortion: Since the sides of the trapezoidal pattern show very little, if any, curvature, it can be said that there is very little, if any, distortion of the modulated wave.

1.20 Determine the modulation factor and percent modulation of the modulated wave which generates the trapezoidal pattern shown as Fig. 1-20.

SOLUTION

Given: Trapezoidal pattern of Fig. 1-20
Find: m and M

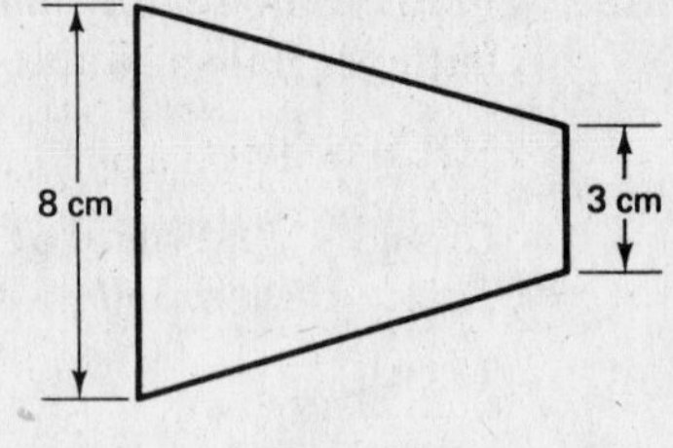

Fig. 1-20

Using

$$M = \frac{x - y}{x + y} \times 100$$

and substituting values from Fig. 1-20, we get

$$M = \frac{8 - 3}{8 + 3} \times 100 = \frac{5}{11} \times 100$$

$$\boxed{M = 45.5\%}$$

The relationship between percent modulation and modulation index is

$$M = m \times 100$$

Substituting numerical values and solving,

$$45.5 = m \times 100$$

$$\frac{45.5}{100} = m$$

$$\boxed{m = 0.455}$$

1.21 An AM standard broadcast receiver is to be designed having an intermediate frequency (IF) of 455 kHz.

(*a*) Calculate the required frequency that the local oscillator should be at when the receiver is tuned to 540 kHz if the local oscillator tracks above the frequency of the received signal.
(*b*) Repeat (*a*) if the local oscillator tracks below the frequency of the received signal.

SOLUTION

Given: $f_{IF} = 455$ kHz
$f_c = 540$ kHz

Find: (*a*) $f_{LO\,\text{above}}$ (*b*) $f_{LO\,\text{below}}$

The intermediate frequency is generated by producing a difference frequency between the carrier and the local oscillator. Thus,

$$f_{IF} = f_c - f_{LO}$$

or

$$f_{IF} = f_{LO} - f_c$$

(*a*)
$$f_{IF} = f_{LO} - f_c$$

Solving for f_{LO},

$$f_{LO} = f_{IF} + f_c$$
$$= (455 \times 10^3) + (540 \times 10^3)$$

$$\boxed{f_{LO} = 995 \text{ kHz}}$$

(*b*)
$$f_{IF} = f_c - f_{LO}$$

Solving for f_{LO},

$$f_{LO} = f_c - f_{IF}$$
$$= (540 \times 10^3) - (455 \times 10^3)$$

$$\boxed{f_{LO} = 85 \text{ kHz}}$$

Supplementary Problems

1.22 Why don't broadcast stations transmit at audio frequencies?
Ans. Stations could not be distinguished from each other. Antennas would have to be tremendously large.

1.23 Which of the AM waves in Fig. 1-21 depicts undermodulation? 100% modulation? overmodulation?
Ans. under: (*a*), (*e*); 100%: (*c*); over: (*b*), (*d*)

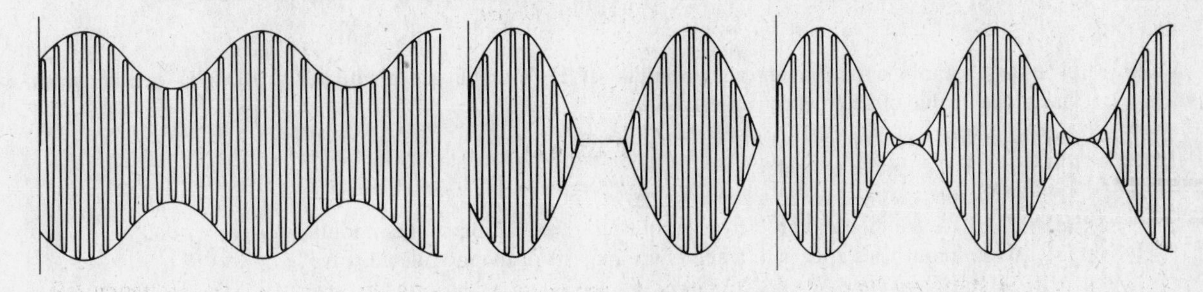

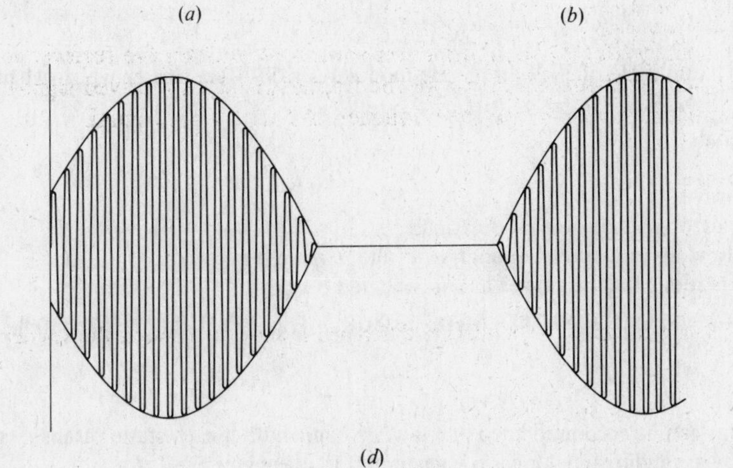

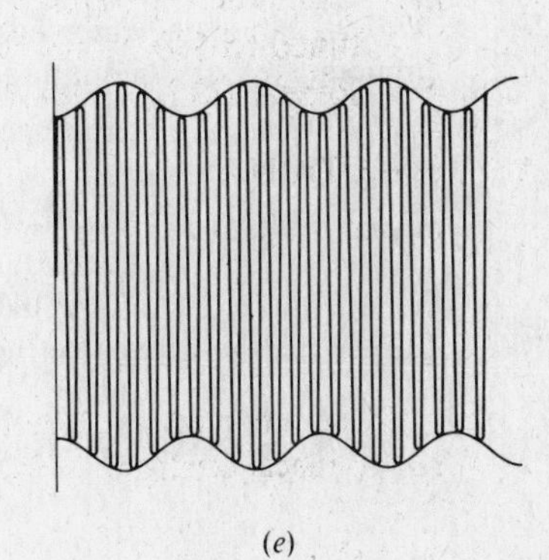

Fig. 1-21

1.24 An audio signal whose mathematical description is

$$25 \sin (2\pi 1000t)$$

modulates a carrier described as

$$75 \sin (2\pi 150\,000t)$$

(*a*) Sketch the audio signal.
(*b*) Sketch the carrier.
(*c*) Construct the modulated wave showing all amplitude magnitudes.
(*d*) Calculate the modulation factor and percent modulation.
(*e*) What is the frequency of the audio signal? of the carrier?
(*f*) What frequencies would show up in a spectrum analysis of the modulated wave?

Ans. (*d*) 0.333, 33.33%; (*e*) 1000 Hz, 150 000 Hz; (*f*) 149 000 Hz, 150 000 Hz, 151 000 Hz

1.25 An audio signal described as

$$30 \sin (2\pi 2500t)$$

amplitude modulates a carrier which is described as

$$65 \sin (2\pi 250\,000t)$$

(*a*) Sketch the audio signal.
(*b*) Sketch the carrier.
(*c*) Construct the modulated wave.
(*d*) What is the modulation factor and percent modulation?
(*e*) What is the frequency of the audio signal? of the carrier?
(*f*) What frequencies would show up in a spectrum analysis of the modulated wave?

Ans. (*d*) 0.4615, 46.15%; (*e*) 2500 Hz, 250 000 Hz; (*f*) 247 500 Hz, 250 000 Hz, 252 500 Hz

1.26 A 2000-Hz audio signal having an amplitude of 15 V amplitude modulates a 100-kHz carrier which has a peak value of 25 V when not modulated.

(*a*) Sketch the audio signal to scale.
(*b*) Sketch the carrier to scale.
(*c*) Construct the modulated wave to scale.
(*d*) Calculate the modulation factor and percent modulation of the modulated wave.
(*e*) What frequencies appear in a spectrum analysis of the modulated wave?

Ans. (*d*) 0.60, 60%; (*e*) 98 kHz, 100 kHz, 102 kHz

1.27 An 1800-Hz signal which has an amplitude of 30 V amplitude modulates a 50-MHz carrier which when unmodulated has an amplitude of 65 V.

(*a*) Sketch the modulating signal.
(*b*) Sketch the carrier.
(*c*) Construct the modulated wave.
(*d*) Calculate the modulation factor and percent modulation.
(*e*) What frequencies would show up in a spectrum analysis of the AM wave?
(*f*) Write the trigonometric equations for the carrier and for the audio signal.

Ans. (*d*) 0.4615, 46.15%; (*e*) 49.9982 MHz, 50.0000 MHz, 50.0018 MHz; (*f*) $65 \sin [2\pi(50 \times 10^6)t]$, $30 \sin [2\pi(1800)t]$

1.28 How many AM broadcast stations can be accommodated in a 6-MHz bandwidth if each station transmits a signal which was modulated by an audio signal having a maximum frequency of 5 kHz?

Ans. 600 stations

1.29 A bandwidth of 12 MHz becomes available for assignment. If assigned for TV broadcast service, only two channels could be accommodated. Determine the number of AM stations that could broadcast simultaneously if the maximum modulating frequency is limited to 5 kHz. *Ans.* 1200 stations

1.30 A 90-kHz bandwidth is to accommodate six AM broadcasts simultaneously. What maximum modulating frequency must each station be limited to? *Ans.* 7500 Hz

1.31 An antenna transmits an AM signal having a total power content of 15 kW. Determine the power being transmitted at the carrier frequency and at each of the sidebands when the percent modulation is 85%. *Ans.* 11 019 W, 1990 W

1.32 Calculate the power content of the carrier and of each of the sidebands of an AM signal whose total broadcast power is 40 kW when the percent modulation is 60%. *Ans.* 33 898 W, 3050 W

1.33 A 3500-Hz audio tone amplitude modulates a 200-kHz carrier resulting in a modulated signal having a percent modulation of 85%. The total power being transmitted is 15 kW.

(*a*) What frequencies would appear in a spectrum analysis of the modulated wave?

(*b*) Determine the power content at each of the frequencies that appear in a spectrum analysis of the modulated wave.

Ans. (*a*) 196 500 Hz, 200 000 Hz, 203 500 Hz; (*b*) 11 019 W, 1990 W

1.34 Determine the power contained at the carrier frequency and within each of the sidebands for an AM signal whose total power content is 15 kW when the modulation factor is 0.70. *Ans.* 12 048 W, 1475 W

1.35 An amplitude-modulated signal contains a total of 6 kW. Calculate the power being transmitted at the carrier frequency and at each of the sidebands when the percent modulation is 100%. *Ans.* 4000 W, 1000 W

1.36 An AM wave has a power content of 1800 W at its carrier frequency. What is the power content of each of the sidebands when the carrier is modulated 85%? *Ans.* 325 W

1.37 An AM signal contains 500 W at its carrier frequency and 100 W in each of its sidebands.

(*a*) Determine the percent modulation of the AM signal.

(*b*) Find the allocation of power if the percent modulation is changed to 60%.

NOTE: The power content of the carrier of an AM wave does not vary with percent modulation.

Ans. (*a*) 89.44%; (*b*) 500 W, 45 W, 45 W

1.38 1200 W is contained at the carrier frequency of an AM signal. Determine the power content of each of the sidebands for each of the following percent modulations: (*a*) 40%, (*b*) 50%, (*c*) 75%, (*d*) 100%. *NOTE*: The power content of the carrier of an AM wave does not vary with percent modulation. *Ans.* (*a*) 48 W, (*b*) 75 W, (*c*) 168.75 W, (*d*) 300 W

1.39 An AM wave has a total transmitted power of 4 kW when modulated 85%. How much total power should an SSB wave contain in order to have the same power content as that contained in the two sidebands? *Ans.* 1061.52 W

1.40 An SSB transmission contains 800 W. This transmission is to be replaced by a standard AM signal with the same power content. Determine the power content of the carrier and each of the sidebands when the percent modulation is 85%. *Ans.* 587.695 W, 106.15 W

1.41 Calculate the modulation factor and percent modulation of the AM wave shown as Fig. 1-22.
Ans. 0.50, 50%

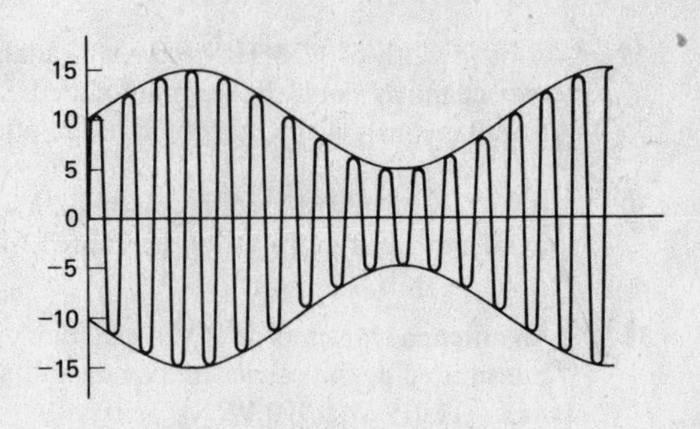

Fig. 1-22

1.42 An AM signal is examined on an oscilloscope. It has a maximum peak to peak of 4.5 cm and a minimum peak to peak of 2 cm.

(*a*) Sketch the pattern observed on the 'scope.
(*b*) Determine the modulation factor and percent modulation of the signal.
(*c*) Calculate the power content of each of the sidebands if the power contained by the signal at the carrier frequency is 500 W.

Ans. (*b*) 0.3846, 38.46%; (*c*) 18.49 W

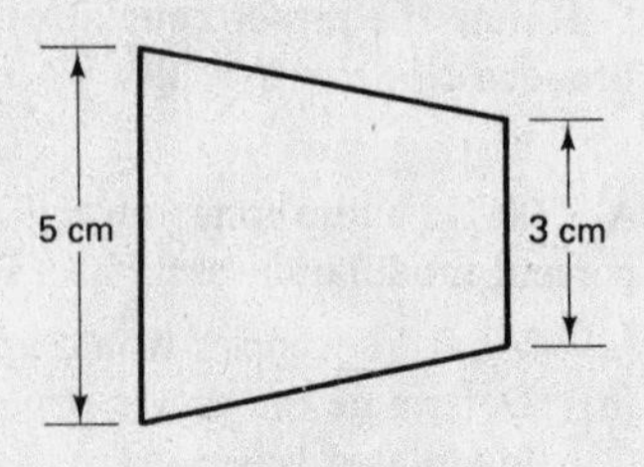

Fig. 1-23

1.43 The trapezoidal pattern shown in Fig. 1-23 results when examining an AM wave. Calculate the modulation factor and percent modulation.

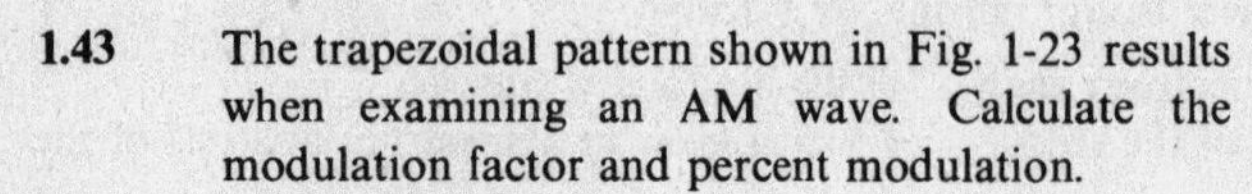

Ans. 0.25, 25%

1.44 Determine the modulation factor and percent modulation of the AM wave which produces the trapezoidal pattern shown in Fig. 1-24. *Ans.* 0.7143, 71.43%

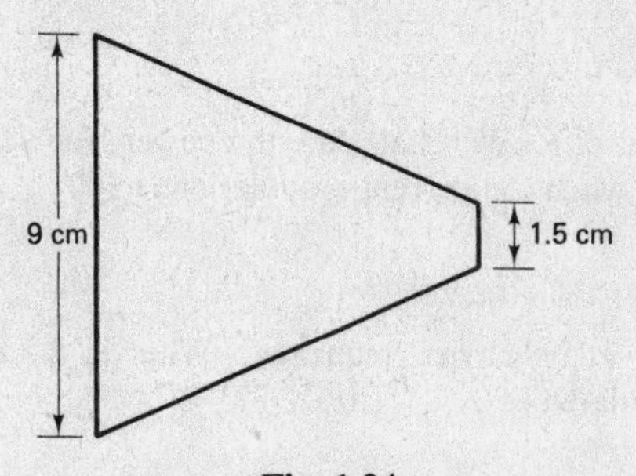

Fig. 1-24

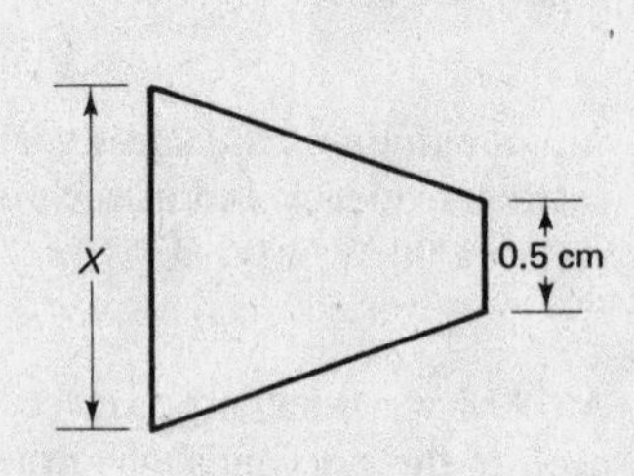

Fig. 1-25

1.45 The trapezoidal pattern of an amplitude-modulated wave having an 85% modulation is shown in Fig. 1-25. Determine the length x of the larger vertical side. *Ans.* 6.167 cm

1.46 Why is it necessary to have a nonlinear device in an AM transmitter?
Ans. To generate sum and difference frequencies.

1.47 Where does modulation occur in most AM transmitters? *Ans.* In the modulated amplifier.

1.48 Why are class C amplifiers suited for use as modulated amplifiers in AM transmitters?
Ans. They are nonlinear.

1.49 What part of an AM transmitter is usually called the modulator?
Ans. The last stage of audio amplification.

1.50 How does single sideband differ from standard AM?
Ans. It is missing the carrier and one sideband.

1.51 How does double sideband differ from standard AM? *Ans.* It is missing the carrier.

1.52 How does pilot-carrier transmission differ from standard AM transmission?
Ans. Only a small portion of the carrier is transmitted.

1.53 What are the advantages of SSB and DSB over AM?
Ans. The power is placed where the information is.

1.54 What is an added advantage of SSB over DSB? *Ans.* Less noise pickup.

1.55 What is a balanced modulator? Why is this a poor choice of name for this device?
Ans. A modulator that puts out only two sidebands and no carrier.

1.56 Why have SSB and DSB not found acceptance for use in most home-type entertainment equipment?
Ans. Cost.

1.57 What function is served by having an intermediate-frequency section in an AM superheterodyne receiver?
Ans. The passband of the receiver stays the same as the receiver is tuned from the lower end of the spectrum to the upper end.

1.58 What is a local oscillator and what is its function?
Ans. An oscillator within the receiver which generates a signal to be heterodyned with the incoming signal.

1.59 An AM receiver is tuned to a station whose carrier frequency is 750 kHz. What frequency should the local oscillator be set to in order to provide an intermediate frequency of 455 kHz if the local oscillator frequency tracks below the received frequency? If it tracks above?
Ans. 295 kHz, 1205 kHz

1.60 Repeat Problem 1.59 for a station having a carrier frequency of 1200 kHz. *Ans.* 745 kHz, 1655 kHz

Chapter 2

Frequency Modulation

Frequency modulation was originally developed to cope with undesirable noise which competed with the desired signal when amplitude modulation was used. Most noise appeared as an additional amplitude modulation on the signal.

When frequency modulating a carrier, information is placed on the carrier by varying its frequency while holding its amplitude fixed. Upon being received, variations in amplitude are eliminated prior to demodulation without affecting the information content contained in the frequency variations, thereby eliminating any noise which may appear as an amplitude modulation of the carrier. See Fig. 2-1.

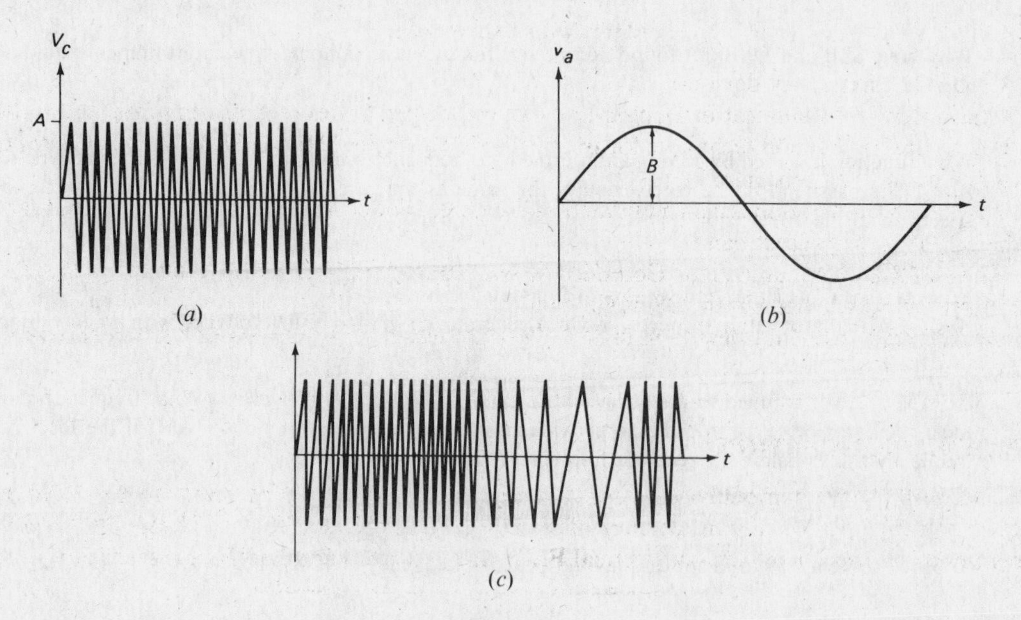

Fig. 2-1

The unmodulated carrier is described as

$$v_c = A \sin 2\pi f t$$

The modulating or audio signal is described as

$$v_a = B \sin 2\pi f_a t$$

The carrier frequency f will vary around a resting frequency f_c thus:

$$f = f_c + \Delta f \sin 2\pi f_a t$$

The frequency-modulated wave will have the following description:

$$v = A \sin [2\pi(f_c + \Delta f \sin 2\pi f_a t)t]$$

In this frequency-modulated situation, Δf is the maximum change in frequency the modulated wave undergoes; it is called the *frequency deviation*. The total variation in frequency, from the lowest to the highest, is referred to as the *carrier swing*. Thus, for a modulating signal which has equal positive and negative peaks, such as a pure sine wave, the carrier swing is equal to two times the frequency deviation.

$$\Delta f = \text{frequency deviation}$$

$$\text{Carrier swing} = 2 \times \text{frequency deviation}$$

It can be shown that the equation for the frequency-modulated wave can be manipulated into

$$v = A \sin\left(2\pi f_c t + \frac{\Delta f}{f_a} \cos 2\pi f_a t\right)$$

However, since the mathematics involved in developing this second equation depends on calculus, it will not be shown.

Note that in this equation the cosine term is preceded by the term $\Delta f/f_a$. This quantity is called the *modulation index* and is indicated as m_f.

$$\text{Modulation index} = m_f = \frac{\Delta f}{f_a}$$

where Δf is the frequency deviation.

The Federal Communications Commission of the United States requires that frequency modulation be used as the modulation technique for the band of frequencies between 88 MHz and 108 MHz. This is known as the *FM broadcast band.*

Frequency modulation is also mandated as the required modulation technique for the audio portion of the television broadcast band.

The Federal Communications Commission sets a maximum frequency deviation of 75 kHz for FM broadcast stations in the 88- to 108-MHz band.

A maximum frequency deviation of 25 kHz is permitted for the sound portion of television broadcasts.

PERCENT MODULATION

The term "percent modulation" as it is used in reference to FM refers to the ratio of actual frequency deviation to the maximum allowable frequency deviation. Thus, 100% modulation corresponds to 75 kHz for the commercial FM broadcast band and 25 kHz for television.

$$\text{Percent modulation } M = \frac{\Delta f_{\text{actual}}}{\Delta f_{\text{max}}} \times 100$$

SIDEBANDS

Analyzing a frequency-modulated wave results in finding that unlike the amplitude-modulated wave, which has only two side frequencies for each modulating frequency, the FM signal has an infinite number of side frequencies spaced f_a apart on both sides of the resting frequency. See Fig. 2-2. Happily, however, most of the side frequencies do not contain significant amounts of power.

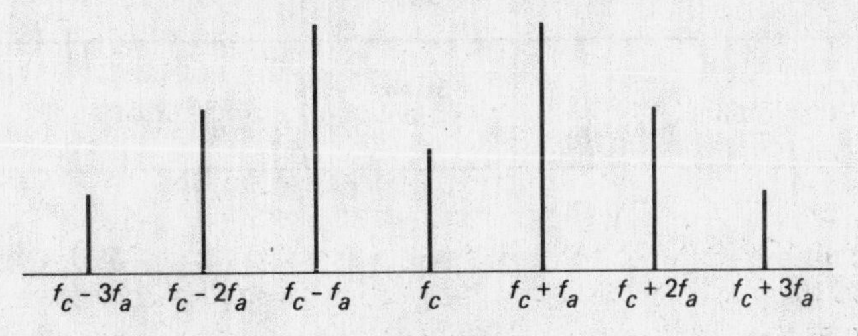

Fig. 2-2

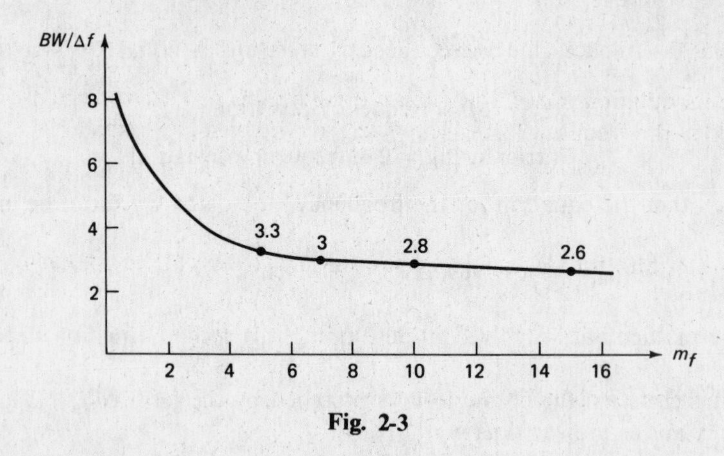

Fig. 2-3

Fourier analysis indicates that the number of side frequencies which contain a significant amount of power, and thus the effective bandwidth of the FM signal, is dependent on the modulation index of the modulated wave, $\Delta f / f_a$.

Schwartz* developed a graph for determining the bandwidth of an FM signal if the modulation index is known. This graph has been reproduced as Fig. 2-3. Schwartz uses as his criterion the rule of thumb that any component frequency with a signal strength (voltage) less than 1% of that of the unmodulated carrier shall be considered too small to be significant.

The Federal Communications Commission sets a 15-kHz maximum limit on the frequency of the modulating signal for both the FM broadcast band and commercial television band (the sound portion of broadcast television transmission is a frequency-modulated signal).

$$f_{a,\,\max} = 15 \text{ kHz} \qquad \text{for both 88 to 108 MHz and TV}$$

CENTER FREQUENCY AND BANDWIDTH ALLOCATIONS

Each commercial FM broadcast station in the 88- to 108-MHz band is allocated a 150-kHz channel plus a 25-kHz guard band at both the upper and the lower edges of the station allocation by the FCC. Thus, a total channel width of 200 kHz is provided to each station in the commercial FM broadcast band.

$$150 \text{ kHz} + 2(25 \text{ kHz}) = 200 \text{ kHz}$$

In addition to this large bandwidth and guard-band combination (200 kHz), only alternate channels are assigned within any particular geographic area. In the UHF band, of which the commercial FM broadcast band is a part, reception is limited to distances only slightly farther than the horizon. Thus, assigning only alternate channels in any given geographic area limits the possibility of interference (see Fig. 2-4).

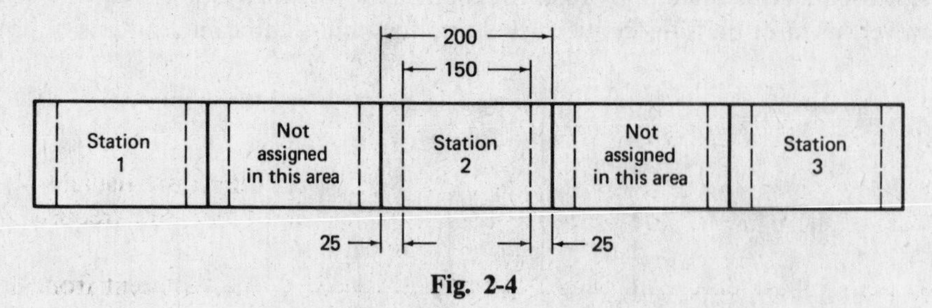

Fig. 2-4

* Mischa Schwartz, *Information Transmission, Modulation, and Noise*, McGraw-Hill, New York, 1959.

DEVIATION RATIO

The worst-case modulation index, in which the maximum permitted frequency deviation and the maximum permitted audio frequency are used, is called the *deviation ratio.*

$$\text{Deviation ratio} = \frac{\Delta f_{\max}}{f_{a,\,\max}}$$

Thus the deviation ratio for stations in the commercial FM broadcast band is

$$\text{Deviation ratio, 88–108 MHz} = \frac{75\text{ kHz}}{15\text{ kHz}} = 5$$

and for the sound portion of commercial television,

$$\text{Deviation ratio, TV} = \frac{25\text{ kHz}}{15\text{ kHz}} = 1.67$$

NARROWBAND FM VERSUS WIDEBAND FM

An examination of the Schwartz bandwidth curve of Fig. 2-3 indicates that at high values of m_f the curve tends towards a horizontal asymptote and at low values of m_f it tends toward the vertical. Detailed mathematical study would indicate that the bandwidth of an FM signal for which m_f is less than $\pi/2$ is dependent mainly upon the frequency of the modulating signal and is quite independent of frequency deviation. Further analysis would show that the bandwidth of an FM signal for which m_f is less than $\pi/2$ is equal to twice the modulating frequency.

$$\text{Bandwidth} = 2f_a \qquad \text{for } m_f < \pi/2$$

Just as with AM, and unlike the situation in which $m_f > \pi/2$, two side frequencies show up for each modulating frequency, one above and one below the frequency of the carrier, each spaced f_a away from the carrier frequency. Because of the limited bandwidth of FM signals with $m_f < \pi/2$, such modulations are referred to as *narrowband FM*, and FM signals with $m_f > \pi/2$ are referred to as *wideband FM*.

Though the spectrum for an AM signal and a narrowband FM signal appear to be the same, a Fourier analysis shows that the magnitude and phase relationships for AM and FM are quite different. See Fig. 2-5 for the frequency spectrum of a narrowband FM signal.

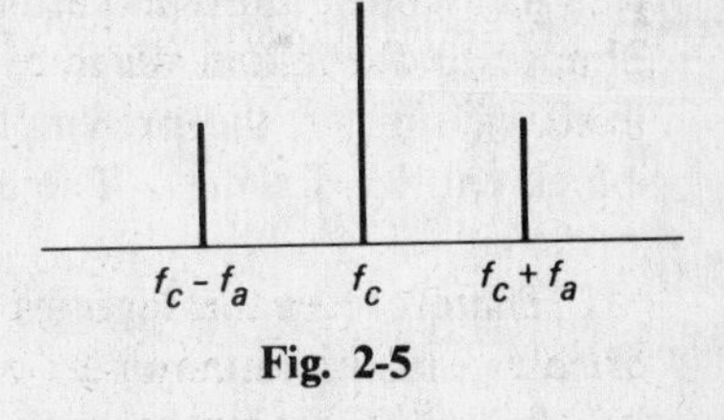

Fig. 2-5

Many of the advantages obtained with wideband FM, such as noise reduction, are not available with narrowband FM. Why, then, would one want to use narrowband FM rather than AM? One reason is that with narrowband FM (as well as with wideband FM) the power content at the carrier frequency decreases as the modulation increases so that we have the desirable situation of putting the power where the information is.

FM RECEIVERS AND TRANSMITTERS

The FM receiver is similar in many ways to the AM receiver. Both are usually superheterodyne receivers. The commercial FM broadcast receiver usually has an intermediate frequency of 10.7 MHz. See Fig. 2-6.

Of course the demodulation circuit for FM receivers will be quite different from that used in an AM receiver. Other differences between the AM and FM receivers are the inclusion of a block called a *limiter* and one called a *de-emphasis network* in the FM receiver.

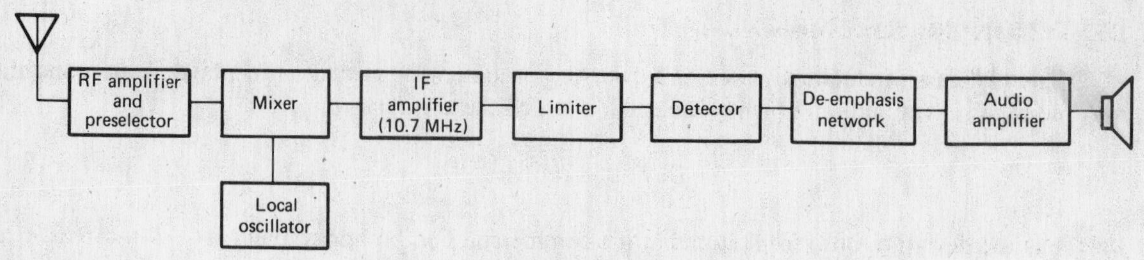

Fig. 2-6

The Limiter

The purpose of the limiter circuit is to clip all amplitude variations which may exist in the signal as it reaches this part of the system. This clipping removes any AM noise which may have become part of the signal. Clipping by the limiter eliminates noise but does not affect the information content of the signal because the information is contained in the frequency variations, not in the amplitude variations.

The De-emphasis Network

The de-emphasis network which appears in the block diagram of the FM receiver is only one-half of a system which consists of both a pre-emphasis and a de-emphasis network, the pre-emphasis network being located in the transmitter. The pre-emphasis network causes the higher-frequency information content of the audio signal at the transmitter to be amplified more than the lower-frequency information. The de-emphasis network compensates for this by reducing the gain of the higher-frequency audio signal. The reason for the inclusion of such a system is to reduce frequency-modulated noise which enters the transmitted signal while en route from the transmitter to the receiver as well as any such noise which may enter at the front end of the receiver.

Investigators found that noise which entered the signal as a frequency modulation occurred with greater likelihood and disturbance in the higher audio frequencies; thus, the pre-emphasis–de-emphasis system functions to reduce frequency-modulated noise.

FM Transmitters

As is most likely expected, the block diagram for an FM transmitter appears to be somewhat similar to the block diagram of an AM transmitter. Note in Fig. 2-7, the block diagram for an FM transmitter, the pre-emphasis network as expected from the discussion of FM receivers and a block entitled Exciter. The exciter is that portion of the FM transmitter within which modulation occurs.

There are two categories of techniques for the generation of an FM signal. One is called the *Direct method* and the other is called the *Indirect method.*

In the Direct method, a tuned circuit containing a device whose capacitance can be made to vary directly with the amplitude of the modulating signal is used. It is placed in shunt with a parallel RLC tank circuit. The most commonly used devices of this sort include the *transistor reactance modulator*, the *reactance tube modulator*, and *varactor diodes* (varicaps).

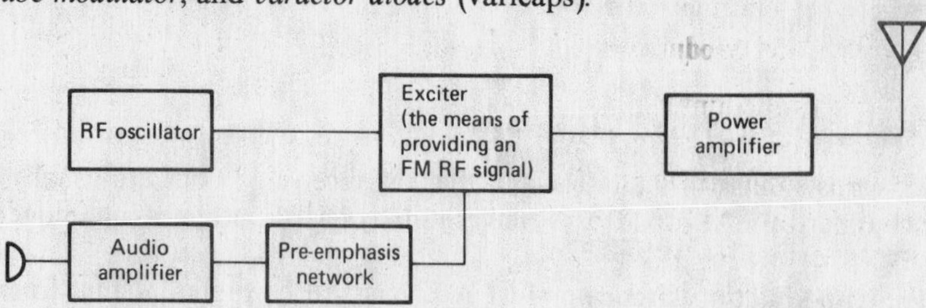

Fig. 2-7

The Transistor Reactance Modulator

Figure 2-8 is a schematic diagram of a transistor reactance modulator. The capacitance presented by this circuit is

$$C_{\text{eq}} = \frac{h_{\text{fe}} R_2 C_2}{h_{\text{ie}} + R_2}$$

The beta (β) of the transistor, h_{fe}, is caused to vary by changing the operating point of the transistor, the operating point being determined by the slowly varying audio signal input.

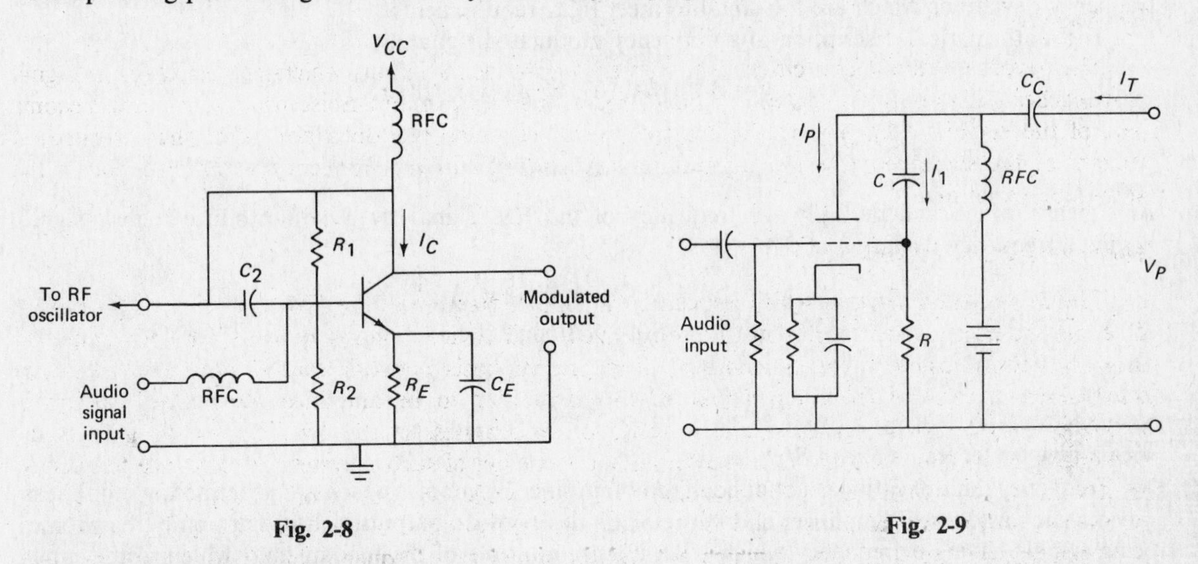

Fig. 2-8

Fig. 2-9

The Reactance Tube Modulator

Figure 2-9 is a schematic diagram of a reactance tube modulator. The capacitance presented by this circuit is

$$C_{\text{eq}} = g_m RC$$

where g_m is caused to vary with audio signal. A remote cutoff tube is usually used because the g_m of this tube is very sensitive to the operating point of the tube.

The Varactor Diode Modulator

Figure 2-10 is a schematic of a circuit which can function as an FM modulator. The varactor diode capacitance varies with its bias, which is determined by V_{CC} and the audio input signal.

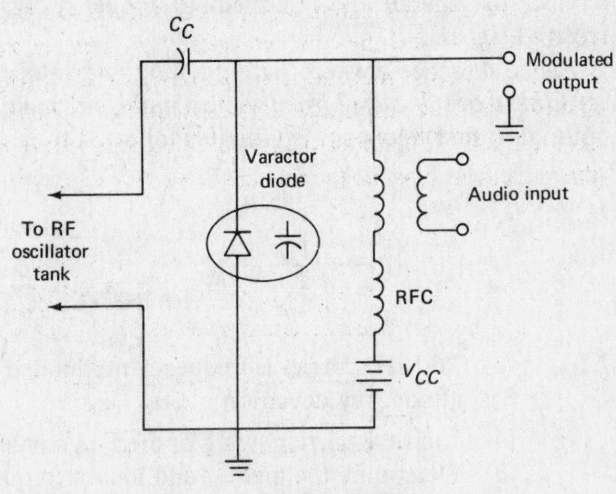

Fig. 2-10

The Indirect Method of FM Modulation

One of the difficulties encountered in FM transmitters which depend upon the Direct method of frequency modulation is that because of the variable nature of the tuning of the tank circuit, crystal-controlled oscillators cannot be used and therefore the stability inherent in such crystal-controlled units is not available.

An alternative technique for the generation of a frequency-modulated signal which permits the use of crystal control is called the Indirect method. In this technique the phase angle is caused to vary while holding the frequency constant. What is really generated by this technique is what is called a *phase-modulated signal.* With some minor doctoring, this phase-modulated signal can be passed off as an FM signal, and it is.

Frequency Multipliers

Most often, FM signals are initially generated in low-power circuits and circuits providing frequency deviations which are too small to meet FCC requirements.

The mathematical description of a frequency-modulated signal is

$$v = A \sin [2\pi(f_c + \Delta f \sin (2\pi f_a t))t]$$

The frequency is

$$f = f_c + \Delta f \sin 2\pi f_a t$$

Any means which will multiply the frequency of the FM signal by S will produce a new signal having a frequency deviation of $S\,\Delta f$.

$$Sf = S(f_c + \Delta f \sin 2\pi f_a t)$$
$$Sf = Sf_c + S\,\Delta f \sin 2\pi f_a t$$

Thus,

$$\Delta f_{\text{new}} = S\,\Delta f_{\text{old}}$$

with a new center frequency of $Sf_{c\,\text{old}}$.

Frequency multiplication is not difficult to obtain since harmonics which are generated by nonlinear devices such as class C amplifiers and varactor diodes provide outputs rich in harmonics, harmonics being signals having a frequency which is an integer multiple of the fundamental which is the input signal. It then becomes merely a chore of choosing the appropriate harmonic by using a frequency-selective circuit.

Frequency multipliers are generally limited by practical considerations to multiplications by 2, 3, or 4. Larger multiplication factors may be obtained by cascading these smaller multipliers.

Heterodyning

It sometimes becomes necessary to be able to adjust the frequency of the modulated signal without affecting frequency deviation. This can be accomplished by mixing, beating, or heterodyning, all three terms meaning the same thing. This is the same process used in the superheterodyne receiver to generate the intermediate frequency by heterodyning the local oscillator signal with the received signal.

The difference between heterodyning and multiplying is that in heterodyning the sinusoid angle is *added to* or *subtracted from*, while a multiplier *multiplies* the sinusoid angle by some factor. It is not unusual to find frequency multipliers followed by a heterodyner in an FM transmitter.

Solved Problems

2.1 A 107.6-MHz carrier is frequency modulated by a 7-kHz sine wave. The resultant FM signal has a frequency deviation of 50 kHz.

(*a*) Find the carrier swing of the FM signal.
(*b*) Determine the highest and lowest frequencies attained by the modulated signal.
(*c*) What is the modulation index of the FM wave?

SOLUTION

Given: $f_c = 107.6$ MHz
$f_a = 7$ kHz
$\Delta f = 50$ kHz

Find: (*a*) c.s. (*b*) f_H, f_L (*c*) m_f

(*a*) Relating carrier swing to frequency deviation,

$$\text{c.s.} = 2\Delta f = 2 \times 50 \times 10^3$$

$$\boxed{\text{c.s.} = 100 \text{ kHz}}$$

(*b*) The upper frequency reached is equal to the rest or carrier frequency plus the frequency deviation:

$$\begin{aligned} f_H &= f_c + \Delta f \\ &= 107.6 \times 10^6 + 50 \times 10^3 \\ &= (107\,600 \times 10^3) + (50 \times 10^3) \\ &= 107\,650 \times 10^3 \end{aligned}$$

$$\boxed{f_H = 107.65 \text{ MHz}}$$

The lowest frequency reached by the modulated wave is equal to the rest or carrier frequency minus the frequency deviation.

$$\begin{aligned} f_L &= f_c - \Delta f \\ &= 107.6 \times 10^6 - 50 \times 10^3 \\ &= 107\,600 \times 10^3 - 50 \times 10^3 \\ &= 107\,550 \times 10^3 \end{aligned}$$

$$\boxed{f_L = 107.55 \text{ MHz}}$$

(*c*) The modulation index is determined by

$$m_f = \frac{\Delta f}{f_a} = \frac{50 \times 10^3}{7 \times 10^3}$$

$$\boxed{m_f = 7.143}$$

2.2 Determine the frequency deviation and carrier swing for a frequency-modulated signal which has a resting frequency of 105.000 MHz and whose upper frequency is 105.007 MHz when modulated by a particular wave. Find the lowest frequency reached by the FM wave.

SOLUTION

Given: $f_0 = 105.000$ MHz
$f_{\text{upper}} = 105.007$ MHz

Find: Δf, c.s., f_{lower}

Frequency deviation is defined as the maximum change in frequency of the modulated signal away from the rest or carrier frequency.

$$\Delta f = (105.007 - 105.000) \times 10^6$$
$$= 0.007 \times 10^6$$
$$= 7000$$

$$\boxed{\Delta f = 7\text{ kHz}}$$

Carrier swing can now be determined by

$$\text{c.s.} = 2\,\Delta f$$
$$= 2(7 \times 10^3)$$
$$= 14 \times 10^3$$

$$\boxed{\text{c.s.} = 14\text{ kHz}}$$

The lowest frequency reached by the modulated wave can be found by subtracting the frequency deviation from the carrier or rest frequency.

$$f_{\text{lower}} = f_0 - \Delta f$$
$$= (105.000 - 0.007) \times 10^6$$

$$\boxed{f_{\text{lower}} = 104.993\text{ MHz}}$$

2.3 What is the modulation index of an FM signal having a carrier swing of 100 kHz when the modulating signal has a frequency of 8 kHz?

SOLUTION

Given: c.s. = 100 kHz

$f_a = 8$ kHz

Find: m_f

From the defining equation,

$$m_f = \frac{\Delta f}{f_a}$$

First determining Δf,

$$\Delta f = \frac{\text{c.s.}}{2}$$
$$= \frac{100 \times 10^3}{2}$$
$$= 50\text{ kHz}$$

Now substituting into the equation for m_f,

$$m_f = \frac{50 \times 10^3}{8 \times 10^3}$$

$$\boxed{m_f = 6.25}$$

2.4 A frequency-modulated signal which is modulated by a 3-kHz sine wave reaches a maximum frequency of 100.02 MHz and minimum frequency of 99.98 MHz.

(*a*) Determine the carrier swing.
(*b*) Find the carrier frequency.
(*c*) Calculate the frequency deviation of the signal.
(*d*) What is the modulation index of the signal?

SOLUTION

Given: $f_{max} = 100.02$ MHz
$f_{min} = 99.98$ MHz
$f_a = 3$ kHz

Find: (*a*) c.s. (*b*) f_c (*c*) Δf (*d*) m_f

(*a*) The carrier swing is defined as the total variation in frequency from the highest to lowest reached by the modulated wave.

$$\begin{aligned} \text{c.s.} &= f_{max} - f_{min} \\ &= 100.02 \times 10^6 - 99.98 \times 10^6 \\ &= 0.04 \times 10^6 \\ &= 40 \times 10^3 \end{aligned}$$

$$\boxed{\text{c.s.} = 40 \text{ kHz}}$$

(*b*) The carrier frequency or rest frequency is midway between the maximum frequency and minimum frequency reached by the modulated wave.

$$\begin{aligned} f_c &= \frac{f_{max} + f_{min}}{2} \\ &= \frac{100.02 \times 10^6 + 99.98 \times 10^6}{2} \\ &= 100 \times 10^6 \end{aligned}$$

$$\boxed{f_c = 100.00 \text{ MHz}}$$

(*c*) Since the carrier swing is equal to twice the frequency deviation,

$$\begin{aligned} \Delta f &= \frac{\text{c.s.}}{2} \\ &= \frac{40 \times 10^3}{2} \end{aligned}$$

$$\boxed{\Delta f = 20 \text{ kHz}}$$

(*d*) The modulation index for a frequency modulated wave is defined as

$$\begin{aligned} m_f &= \frac{\Delta f}{f_a} \\ &= \frac{20 \times 10^3}{3 \times 10^3} \end{aligned}$$

$$\boxed{m_f = 6.667}$$

2.5 An FM transmission has a frequency deviation of 20 kHz.

(*a*) Determine the percent modulation of this signal if it is broadcast in the 88–108 MHz band.
(*b*) Calculate the percent modulation if this signal were broadcast as the audio portion of a television broadcast.

SOLUTION

Given: $\Delta f = 20\ \text{kHz}$

Find: (*a*) Percent modulation—FM broadcast band
(*b*) Percent modulation—TV

(*a*) Percent modulation for an FM wave is defined as

$$M = \frac{\Delta f_{\text{actual}}}{\Delta f_{\text{max}}} \times 100$$

The maximum frequency deviation in the FM broadcast band permitted by the FCC is 75 kHz:

$$M = \frac{20 \times 10^3}{75 \times 10^3} \times 100$$

$$\boxed{M = 26.67\%}$$

(*b*)

$$M = \frac{\Delta f_{\text{actual}}}{\Delta f_{\text{max}}} \times 100$$

The maximum frequency deviation for the FM audio portion of a TV broadcast is 25 kHz as set by the FCC.

$$M = \frac{20 \times 10^3}{25 \times 10^3} \times 100$$

$$\boxed{M = 80.0\%}$$

2.6 (*a*) What is the frequency deviation and carrier swing necessary to provide 75% modulation in the FM broadcast band?
(*b*) Repeat for an FM signal serving as the audio portion of a TV broadcast.

SOLUTION

Given: $M = 75\%$

Find: (*a*) Δf_{FM}, c.s.$_{\text{FM}}$ (*b*) Δf_{TV}, c.s.$_{\text{TV}}$

(*a*) Frequency deviation is defined as

$$M = \frac{\Delta f_{\text{actual}}}{\Delta f_{\text{max}}} \times 100$$

The maximum frequency deviation permitted in the FM broadcast band, 88–108 MHz, by the FCC is 75 kHz.

$$75 = \frac{\Delta f_{\text{FM}}}{75 \times 10^3} \times 100$$

$$\Delta f_{\text{FM}} = \frac{75 \times 75 \times 10^3}{100}$$

$$= 56.25 \times 10^3$$

$$\boxed{\Delta f_{\text{FM}} = 56.25\ \text{kHz}}$$

Carrier swing is related to frequency deviation by

$$\text{c.s.}_{FM} = 2\Delta f_{FM}$$
$$= 2 \times 56.25 \times 10^3$$

$$\boxed{\text{c.s.}_{FM} = 112.5 \text{ kHz}}$$

(*b*)
$$M = \frac{\Delta f_{actual}}{\Delta f_{max}} \times 100$$

The maximum frequency deviation permitted by the FCC for the audio portion of a TV signal is 25 kHz. Thus,

$$75 = \frac{\Delta f_{TV}}{25 \times 10^3} \times 100$$

$$\Delta f_{TV} = \frac{75 \times 25 \times 10^3}{100}$$

$$\boxed{\Delta f_{TV} = 18.75 \text{ kHz}}$$

$$\text{c.s.}_{TV} = 2\Delta f_{TV}$$
$$= 2 \times 18.75 \times 10^3$$

$$\boxed{\text{c.s.}_{TV} = 37.5 \text{ kHz}}$$

2.7 Determine the percent modulation of an FM signal which is being broadcast in the 88–108 MHz band, having a carrier swing of 125 kHz.

SOLUTION

Given: c.s. = 125 kHz

Find: M

Frequency deviation and carrier swing are related by

$$\Delta f = \frac{\text{c.s.}}{2}$$
$$= \frac{125 \times 10^3}{2}$$
$$= 62.5 \text{ kHz}$$

$$M = \frac{\Delta f_{actual}}{\Delta f_{max}} \times 100$$

Maximum frequency deviation for the FM broadcast band permitted by the FCC is 75 kHz:

$$M = \frac{62.5 \times 10^3}{75 \times 10^3} \times 100$$

$$\boxed{M = 83.3\%}$$

2.8 The percent modulation of the sound portion of a TV signal is 80%. Determine the frequency deviation and carrier swing of the signal.

SOLUTION

Given: $M = 80\%$

Find: Δf, c.s.

The percent modulation of an FM signal is

$$M = \frac{\Delta f_{\text{actual}}}{\Delta f_{\text{max}}} \times 100$$

The maximum frequency deviation for the sound portion of a TV signal as specified by the FCC is 25 kHz. Thus,

$$80 = \frac{\Delta f_{\text{actual}}}{25 \times 10^3} \times 100$$

$$\Delta f_{\text{actual}} = \frac{80 \times 25 \times 10^3}{100}$$

$$\boxed{\Delta f_{\text{actual}} = 20 \text{ kHz}}$$

Carrier swing is related to frequency deviation by

$$\text{c.s.} = 2\Delta f_{\text{actual}}$$
$$= 2 \times 20 \times 10^3$$

$$\boxed{\text{c.s.} = 40 \text{ kHz}}$$

2.9 A 5-kHz audio tone is used to modulate a 50-MHz carrier causing a frequency deviation of 20 kHz. Determine (*a*) the modulation index and (*b*) the bandwidth of the FM signal.

SOLUTION

Given: $f_a = 5$ kHz
$f_c = 50.0$ MHz
$\Delta f = 20$ kHz

Find: (*a*) m_f (*b*) BW

(*a*) Modulation index is defined as

$$m_f = \frac{\Delta f}{f_a}$$
$$= \frac{20 \times 10^3}{5 \times 10^3}$$

$$\boxed{m_f = 4}$$

(*b*) Referring to the Schwartz bandwidth curve, Fig. 2-3, and entering on the horizontal axis with $m_f = 4$, it is found that

$$\frac{\text{BW}}{\Delta f} = 3.8$$

This is shown in Fig. 2-11.

Substituting 20×10^3 for Δf as given,

$$\frac{\text{BW}}{20 \times 10^3} = 3.8$$

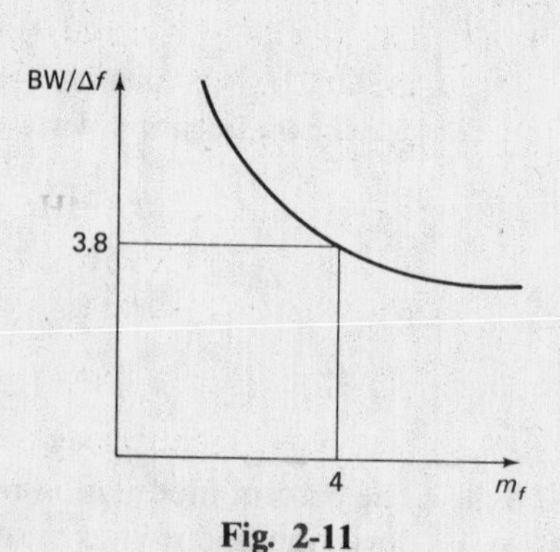

Fig. 2-11

Solving for BW,

$$BW = 3.8 \times 20 \times 10^3$$
$$= 76 \times 10^3$$

$$\boxed{BW = 76 \text{ kHz}}$$

2.10 Determine the frequency of the modulating signal which is producing an FM signal having a bandwidth of 50 kHz when the frequency deviation of the FM signal is 10 kHz.

SOLUTION

Given: $BW = 50$ kHz
$\Delta f = 10$ kHz

Find: f_a

In order to find f_a, reference must be made to the Schwartz bandwidth curve, Fig. 2-3. In order to enter this curve, determine $BW/\Delta f$:

$$\frac{BW}{\Delta f} = \frac{50 \times 10^3}{10 \times 10^3}$$
$$= 5$$

From Fig. 2-3,

$$m_f = 2$$
$$= \frac{\Delta f}{f_a}$$

So,

$$2 = \frac{10 \times 10^3}{f_a}$$
$$f_a = \frac{10 \times 10^3}{2}$$

$$\boxed{f_a = 5 \text{ kHz}}$$

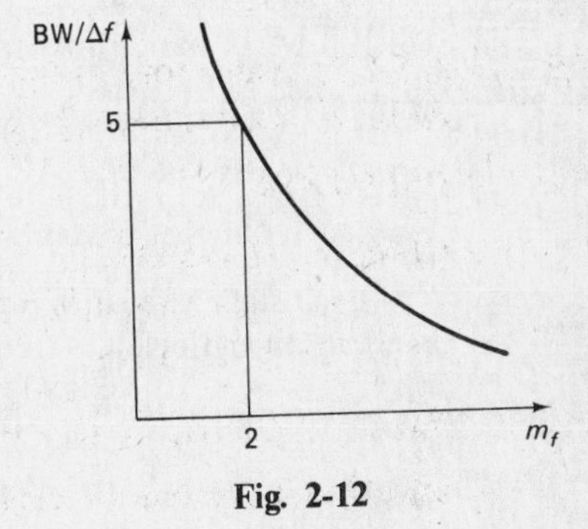

Fig. 2-12

This is shown in Fig. 2-12.

2.11 A 103.0-MHz carrier is frequency modulated by a 10-kHz sine wave. Determine the modulation index of the FM signal. Referring to Schwartz's curve, Fig. 2-3, determine the bandwidth when the carrier swing is 80 kHz.

SOLUTION

Given: $f_c = 103.0$ MHz
$f_a = 10$ kHz
c.s. = 80 kHz

Find: m_f, BW

The defining equation for the modulation index is

$$m_f = \frac{\Delta f}{f_a}$$

However, before using this equation, it is necessary to determine the frequency deviation, Δf.

$$\Delta f = \frac{\text{c.s.}}{2}$$

$$= \frac{80 \text{ kHz}}{2}$$

$$= 40 \text{ kHz}$$

Returning to the defining equation for modulation index:

$$m_f = \frac{40 \times 10^3}{10 \times 10^3}$$

$$\boxed{m_f = 4}$$

Entering the Schwartz bandwidth curve of Fig. 2-3 with $m_f = 4$ results in

$$\frac{\text{BW}}{\Delta f} = 3.5$$

$$\text{BW} = 3.5 \times 40 \times 10^3$$

$$= 140 \times 10^3$$

$$\boxed{\text{BW} = 140 \text{ kHz}}$$

2.12 If a 6-MHz band were being considered for use with the same standards that apply to the 88–108 MHz band, how many FM stations could be accommodated?

SOLUTION

Given: BW = 6 MHz

Find: Number of stations

Each station requires a total bandwidth of 400 kHz; 150 kHz for the signal and a 25-kHz guard band above and below with only alternate channels used.

$$\text{Number of stations} = \frac{6 \times 10^6}{400 \times 10^3}$$

$$\boxed{\text{Number of stations} = 15}$$

2.13 Determine the bandwidth of a narrowband FM signal which is generated by a 4-kHz audio signal modulating a 125-MHz carrier.

SOLUTION

Given: Narrowband FM

$f_a = 4$ kHz

$f_c = 125$ MHz

Find: BW

Since this is a narrowband FM signal, the bandwidth is found merely by doubling the modulating frequency:

$$\text{BW} = 2f_a$$

$$= 2 \times 4 \times 10^3$$

$$\boxed{\text{BW} = 8 \text{ kHz}}$$

2.14 A 2-kHz audio signal modulates a 50-MHz carrier, causing a frequency deviation of 2.5 kHz. Determine the bandwidth of the FM signal.

SOLUTION

Given: $f_a = 2$ kHz
$f_c = 50$ MHz
$\Delta f = 2.5$ kHz

Find: BW

$$m_f = \frac{\Delta f}{f_a} = \frac{2.5 \times 10^3}{2 \times 10^3} = 1.25$$

Since this is less than $\pi/2$, we are dealing with a narrowband signal; thus,

$$\text{BW} = 2f_a$$

$$\boxed{\text{BW} = 2 \times 2 \times 10^3 = 4 \text{ kHz}}$$

2.15 The transistor reactance modulator shown in Fig. 2-13 is to be used in an FM transmitter. The input resistance of the transistor h_{ie} is 600 Ω and the beta of the transistor is 65.

(*a*) How much capacitance does this circuit present to the tank circuit it is attached to?

(*b*) If the beta of the transistor can be caused to swing from 50 to 75, determine the swing in capacitance presented by this circuit.

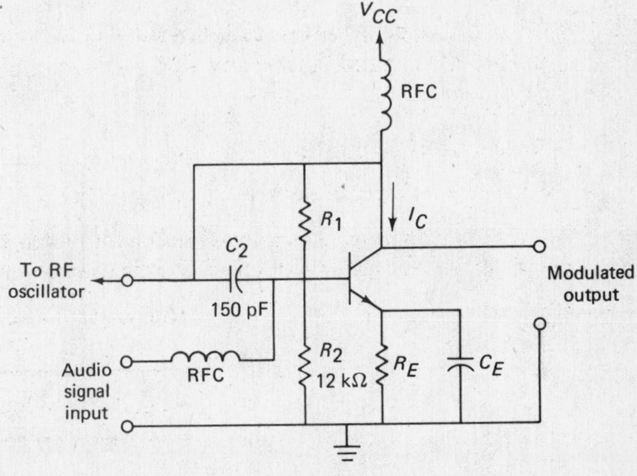

Fig. 2-13

SOLUTION

Given: $h_{ie} = 600\ \Omega$
$\beta_0 = 65$
$\beta_1 = 50$
$\beta_2 = 75$

Find: (*a*) C_{eq2} (*b*) C_{eq1}, C_{eq2}

(*a*) The equation for determining the equivalent capacitance of a transistor reactance modulator is

$$C_{eq} = \frac{h_{fe} R_2 C_2}{h_{ie} + R_2}, \qquad \text{where } h_{fe} = \beta$$

Substituting (values for C_2 and R_2 are taken from Fig. 2-13),

$$C_{eq} = \frac{65(12 \times 10^3)(150 \times 10^{-12})}{600 + 12\,000} = 9.29 \times 10^{-9}$$

$$\boxed{C_{eq} = 9.29 \text{ nF or } 9290 \text{ pF}}$$

(*b*) Using the equation for equivalent capacitance of the transistor reactance modulator,

$$C_{eq\,1} = \frac{h_{fe\,1} R_2 C_2}{h_{ie} + R_2}$$

and substituting numerical values,

$$C_{eq\,1} = \frac{(50)(12 \times 10^3)(150 \times 10^{-12})}{600 + 12\,000}$$
$$= 7.143 \times 10^{-9}$$

The lower value of equivalent capacitance reached is therefore

$$\boxed{C_{eq\,1} = 7.14 \text{ nF} \quad \text{or} \quad 7140 \text{ pF}}$$

Again using

$$C_{eq\,2} = \frac{h_{fe\,2} R_2 C_2}{h_{ie} + R_2}$$

and substituting appropriate numerical values,

$$C_{eq\,2} = \frac{(75)(12 \times 10^3)(150 \times 10^{-12})}{600 + 12\,000}$$

Thus the higher value reached is

$$C_{eq\,2} = 10.71 \times 10^{-9}$$

So,

$$\boxed{C_{eq\,2} = 10.71 \text{ nF} \quad \text{or} \quad = 10\,710 \text{ pF} \quad \text{or} \quad = 0.010\,71\ \mu\text{F}}$$

2.16 The reactance tube modulator shown in Fig. 2-14 uses a remote cutoff tube whose transconductance g_m varies from 2500 μS to 3500 μS (in the SI metric system, the unit of conductance is the siemens, abbreviated S). Determine the range of capacitance it presents.

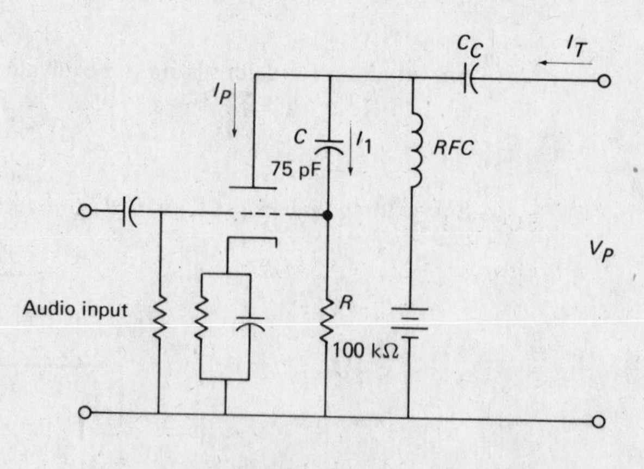

Fig. 2-14

SOLUTION

Given:
$g_{m1} = 2500\ \mu$S
$g_{m2} = 3500\ \mu$S
$C = 75$ pF (from Fig. 2-14)
$R = 100$ kΩ (from Fig. 2-14)

Find: $C_{eq\,1}$, $C_{eq\,2}$

The appropriate formula for the equivalent capacitance of a reactance tube modulator is

$$C_{eq} = g_m RC$$

Using the lower value of g_m,

$$C_{eq1} = (2500 \times 10^{-6})(100 \times 10^3)(75 \times 10^{-12})$$
$$= 18.75 \times 10^{-9}$$

Thus the lower value reached by this reactance tube modulator is

$$\boxed{\begin{aligned} C_{eq1} &= 18.75 \text{ nF} \\ &\text{or} \\ &= 18\,750 \text{ pF} \\ &\text{or} \\ &= 0.018\,75\ \mu\text{F} \end{aligned}}$$

Using the same formula to determine the high value reached by C_{eq},

$$C_{eq2} = g_{m2} RC$$
$$= (3500 \times 10^{-6})(100 \times 10^3)(75 \times 10^{-12})$$
$$= 26.25 \times 10^{-9}$$

Thus the highest value reached by C_{eq} is

$$\boxed{\begin{aligned} C_{eq2} &= 26.25 \text{ nF} \\ &\text{or} \\ &= 26\,250 \text{ pF} \\ &\text{or} \\ &= 0.026\,25\ \mu\text{F} \end{aligned}}$$

2.17 Figure 2-15 is the block diagram of the frequency multiplier and heterodyne portion of an FM transmitter. Calculate the carrier frequency and frequency deviation of each of the points: (*a*) 1, (*b*) 2, and (*c*) 3.

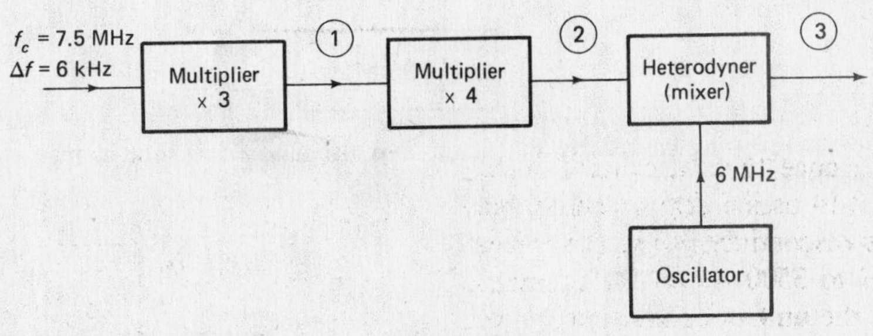

Fig. 2-15

SOLUTION

Given: $f_{c0} = 7.5$ MHz
$\Delta f_0 = 6$ kHz
$S_1 = 3$
$S_2 = 4$
$f_{osc} = 6$ MHz

Find: (*a*) $f_{c1}, \Delta f_1$ (*b*) $f_{c2}, \Delta f_2$ (*c*) $f_{c3}, \Delta f_3$

(*a*) The effect of the first multiplier is to multiply both carrier frequency and frequency deviation by 3.

$$f_{c1} = S_1 f_{co}$$
$$= 3(7.5 \times 10^6)$$
$$= 22.5 \times 10^6$$

$$\boxed{f_{c1} = 22.5 \text{ MHz}}$$

$$\Delta f_1 = S \, \Delta f_0$$
$$= 3(6 \times 10^3)$$
$$= 18 \times 10^3$$

$$\boxed{\Delta f_1 = 18 \text{ kHz}}$$

(*b*) The effect of the second multiplier is to multiply the carrier frequency and frequency deviation present at its input by 4.

$$f_{c2} = S_2 f_{c1}$$
$$= 4(22.5 \times 10^6)$$

$$\boxed{f_{c2} = 90 \text{ MHz}}$$

$$\Delta f_2 = S_2 \, \Delta f_1$$
$$= 4(18 \times 10^3)$$
$$= 72 \times 10^3$$

$$\boxed{\Delta f_2 = 72 \text{ kHz}}$$

(*c*) The mixer has the effect of raising or lowering the carrier frequency by an amount equal to the oscillator frequency. Assuming a rise in frequency,

$$f_{c3} = f_{c2} + f_{osc}$$
$$= (90 \times 10^6) + (6 \times 10^6)$$
$$= 96 \times 10^6$$

$$\boxed{f_{c3} = 96 \text{ MHz}}$$

The heterodyning caused by the mixer and oscillator does not change the frequency deviation. Therefore

$$\Delta f_3 = \Delta f_2$$

$$\boxed{\Delta f_3 = 72 \text{ kHz}}$$

2.18 A 50-MHz FM signal is desired which is to have a frequency deviation of 24 kHz. The output of early stages of the transmitter is a 5-MHz signal with a frequency deviation of 4 kHz. How can the desired output be obtained? See Fig. 2-16.

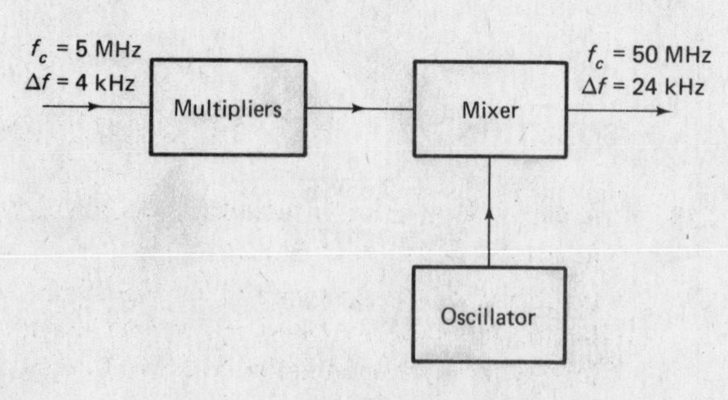

Fig. 2-16

SOLUTION

Given: $f_{c\,\text{out}} = 50$ MHz
$\Delta f_{\text{out}} = 24$ kHz
$f_{c\,\text{in}} = 5$ MHz
$\Delta f_{\text{in}} = 4$ kHz

Find: System

First, find the amount of frequency multiplication required in order to provide the desired frequency deviation.

$$\Delta f_{\text{out}} = S\,\Delta f_{\text{in}}$$
$$24 \times 10^3 = S(4 \times 10^3)$$
$$S = \frac{24 \times 10^3}{4 \times 10^3}$$
$$\boxed{S = 6}$$

Since it is most convenient to specify frequency multipliers which multiply by 2, 3, or 4, a cascade arrangement of a 2× multiplier and a 3× multiplier is required.

This frequency multiplication, although it provides an appropriate value of frequency deviation, doesn't necessarily cause the carrier frequency to have the required value. The carrier frequency obtained by passing the signal through the multipliers is found from

$$f_{c2} = Sf_{c1}$$
$$= 6(5 \times 10^6)$$
$$= 30 \times 10^6$$
$$\boxed{f_{c2} = 30 \text{ MHz}}$$

This frequency of carrier can be caused to change without affecting the frequency deviation by passing the signal through a heterodyne section made up of a mixer and oscillator. The frequency of the oscillator can be found as follows:

$$f_{c2} + f_{\text{osc}} = f_{c3}$$
$$30 \times 10^6 + f_{\text{osc}} = 50 \times 10^6$$
$$f_{\text{osc}} = (50 \times 10^6) - (30 \times 10^6)$$
$$= 20 \times 10^6$$
$$\boxed{f_{\text{osc}} = 20 \text{ MHz}}$$

Supplementary Problems

2.19 A 93.2-MHz carrier is frequency modulated by a 5-kHz sine wave. The resultant FM signal has a frequency deviation of 40 kHz.

(*a*) Find the carrier swing of the FM signal.
(*b*) Determine the highest and lowest frequencies attained by the modulated signal.
(*c*) What is the modulation index of the FM wave?

Ans. (*a*) 80 kHz, (*b*) 93.16 MHz, 93.24 MHz, (*c*) 8

2.20 Determine the carrier swing, the highest and lowest frequencies attained, and the modulation index of the FM signal generated by frequency modulating a 101.6-MHz carrier with an 8-kHz sine wave causing a frequency deviation of 40 kHz. *Ans.* 80 kHz; 101.64 MHz, 101.56 MHz; 5

2.21 Calculate the frequency deviation and carrier swing of a frequency-modulated wave which was produced by modulating a 50.400 MHz carrier. The highest frequency reached by the FM wave is 50.406 MHz. Then calculate the lowest frequency reached by the FM wave.
Ans. 6 kHz, 12 kHz, 50.394 MHz

2.22 Find the upper and lower frequencies that are reached by a frequency-modulated wave that has a rest frequency of 104.003 MHz and a frequency deviation of 60 kHz. What is the carrier swing of the modulated signal? *Ans.* 104.063 MHz, 103.943 MHz, 120 kHz

2.23 Determine the frequency deviation and carrier swing for a frequency-modulated signal which has a resting frequency of 97.340 MHz and whose upper frequency is 97.350 MHz when modulated by a particular audio sine wave. Find the lower frequency reached by the FM wave.
Ans. 10 kHz, 20 kHz, 97.330 MHz

2.24 The carrier swing of a frequency-modulated signal is 120 kHz. The modulating signal is a 6-kHz sine wave. Determine the modulation index of the FM signal. *Ans.* 10

2.25 Determine the modulation index of a frequency-modulated signal having a frequency deviation of 70 kHz. The modulating signal has a frequency of 10 kHz. *Ans.* 7

2.26 A 12-kHz sine wave is to frequency modulate a carrier causing a carrier swing of 80 kHz. Determine the modulation index. *Ans.* 3.333

2.27 Determine the carrier swing, carrier frequency, frequency deviation, and modulation index for a frequency-modulated signal which reaches a maximum frequency of 99.047 MHz and a minimum frequency of 99.023 MHz. The frequency of the modulating signal is 7 kHz.
Ans. 24 kHz, 99.035 MHz, 12 kHz, 1.7

2.28 A carrier is frequency modulated by a 4-kHz sine wave resulting in an FM signal having a maximum frequency of 107.218 MHz and a minimum frequency of 107.196 MHz.

(*a*) Find the carrier swing.
(*b*) Calculate the carrier frequency.
(*c*) What is the frequency deviation of the FM signal?
(*d*) Determine the modulation index of the FM signal.

Ans. (*a*) 22 kHz, (*b*) 107.207 MHz, (*c*) 11 kHz, (*d*) 2.75

2.29 A frequency-modulated signal reaches a maximum frequency of 64.073 MHz and a minimum frequency of 64.050 MHz when modulated by a 4.5-kHz sine wave. Determine the carrier swing, the carrier frequency, the frequency deviation, and the modulation index of the FM signal.
Ans. 23 kHz, 64.0615 MHz, 11.5 kHz, 2.56

2.30 An FM signal for broadcast in the 88–108 MHz band has a frequency deviation of 15 kHz. Find the percent modulation of this signal. If this signal were prepared for broadcast as the audio portion of a television program, what would the percent modulation be? *Ans.* 20%, 60%

2.31 The audio portion of a TV broadcast is to be modulated 80%. Determine the resultant frequency deviation and carrier swing. *Ans.* 20 kHz, 40 kHz

2.32 An FM signal to be broadcast in the 88–108 MHz FM broadcast band is to be modulated 70% Determine the frequency deviation and carrier swing. *Ans.* 52.5 kHz, 105 kHz

2.33 Calculate the frequency deviation and carrier swing necessary to provide an 80% modulation in the FM broadcast band. Repeat this for an FM signal serving as the audio portion of a TV broadcast. *Ans.* 60 kHz, 120 kHz; 20 kHz, 40 kHz

2.34 Find the percent modulation of an FM signal which is being broadcast in the 88–108 MHz band having a carrier swing of 150 kHz. *Ans.* 100%

2.35 Determine the frequency deviation and carrier swing of an FM signal which is the audio portion of a TV signal and has a percent modulation of 85%. *Ans.* 21.25 kHz, 42.5 kHz

2.36 An 8-kHz audio tone is used to modulate a 50.0-MHz carrier causing a frequency deviation of 20 kHz. Determine the modulation index and the bandwidth of the FM signal. *Ans.* 2.5, 100 kHz

2.37 Calculate the modulation index and bandwidth of an FM signal having a 50-kHz frequency deviation when modulated by a 7-kHz audio tone. *Ans.* 7.14, 150 kHz

2.38 Find the modulation index and bandwidth of an FM signal generated by modulating a 100.0-MHz carrier with a 3.8-kHz audio tone causing a carrier swing of 60 kHz. *Ans.* 7.89, 87 kHz

2.39 Determine the frequency of the modulating signal which is producing an FM signal having a bandwidth of 60 kHz when the frequency deviation of the FM signal is 20 kHz. *Ans.* 2.857 kHz

2.40 An FM signal has a bandwidth of 100 kHz when its frequency deviation is 25 kHz. Find the frequency of the modulating signal. *Ans.* 7.1 kHz

2.41 Find the frequency of the audio signal which is frequency modulating a 100-MHz carrier causing a frequency deviation of 25 kHz resulting in an FM signal having a bandwidth of 80 kHz. *Ans.* 4.5 kHz

2.42 An FM signal has a frequency deviation of 40 kHz when the bandwidth of the signal is 160 kHz. Determine the frequency of the modulating signal. *Ans.* 11.4 kHz

2.43 A 15-kHz sine wave is frequency modulating a 104.500-MHz carrier. Determine the modulation index of the FM signal and determine the bandwidth of the FM signal if the carrier swing is 130 kHz. *Ans.* 214.5 kHz

2.44 If an 18-MHz band were to be considered for use with the same standards that apply to the 88–108 MHz FM broadcast band, how many FM stations could be accommodated? *Ans.* 45

2.45 What is the bandwidth of a narrowband FM signal which is generated by a 5-kHz audio signal modulating a 115-MHz carrier? *Ans.* 10 kHz

2.46 A 50.004-MHz carrier is to be frequency modulated by a 3-kHz audio tone resulting in a narrowband FM signal. Determine the bandwidth of the FM signal. *Ans.* 6 kHz

2.47 Determine the bandwidth of a signal generated by a 2.8-kHz audio tone frequency modulating a 98.004-MHz carrier resulting in a frequency deviation of 3.00 kHz. *Ans.* 5.6 kHz

2.48 A frequency deviation of 4 kHz results from frequency modulating a 106.00-MHz carrier. The modulating signal is a 3500-Hz sine wave. Determine the bandwidth of the FM signal. *Ans.* 7000 Hz

2.49 What function is served by the limiter in an FM receiver? *Ans.* To remove amplitude variations.

2.50 Why does the limiter not affect the information content of the signal?
Ans. The information is contained in the frequency variations.

2.51 How does the function of the pre-emphasis–de-emphasis network differ from the function of the limiter circuits? *Ans.* Pre-emphasis–de-emphasis system removes FM noise. Limiter removes AM noise.

2.52 What is an exciter? *Ans.* The means of providing an FM signal.

2.53 How does the Direct method of frequency modulation differ from the Indirect method?
Ans. In the Indirect method, a phase-modulated signal is first generated.

2.54 What is a varicap? What is its function in the exciter section of an FM transmitter?
Ans. A capacitor whose capacitance varies with applied voltage. Used in variable oscillators.

2.55 The input resistance of the transistor shown as part of the transistor reactance modulator of Fig. 2-17 is 450 Ω while the beta of the transistor is 80. What value of capacitance does this circuit present to the tank it is placed in shunt with? If the beta of the transistor swings between 60 and 100, what are the lower and upper values of capacitance presented by this circuit?
Ans. 39.11 nF; 29.33 nF, 48.89 nF

2.56 Repeat Problem 2.55 if the circuit component values are changed to $C_2 = 400$ pF and $R_2 = 8$ kΩ. The same transistor is used.
Ans. 30.28 nF; 22.71 nF, 37.86 nF

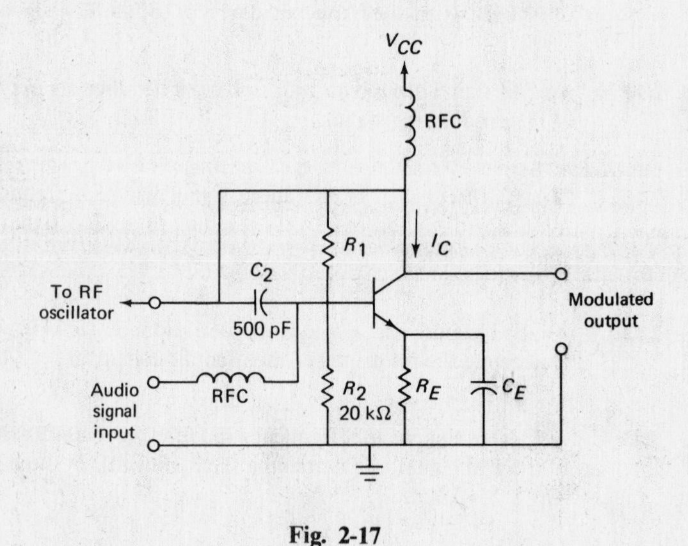

Fig. 2-17

2.57 What new values of equivalent capacitance are presented by the circuit of Fig. 2-17 if the transistor is replaced with one whose beta is 90 and swings between 75 and 105? The input resistance of the replacement transistor is 1500 Ω. *Ans.* 41.85 nF, 34.88 nF, 48.83 nF

2.58 R_2 and C_2 of Fig. 2-17 are replaced with a resistor of 16 kΩ and a capacitor of 150 pF. The transistor to be used with this reactance modulator has a beta that swings between 35 and 60. The transistor has an input resistance of 1200 Ω. Calculate the lower and upper values of capacitance that are reached by this modulator. *Ans.* 4.88 nF, 8.37 nF

2.59 The transconductance of the vacuum tube used in the reactance tube modulator of Fig. 2-18 is made to vary from 2800 μS to 4300 μS. Calculate the range of equivalent capacitance it presents.
Ans. 25.2 nF, 38.7 nF

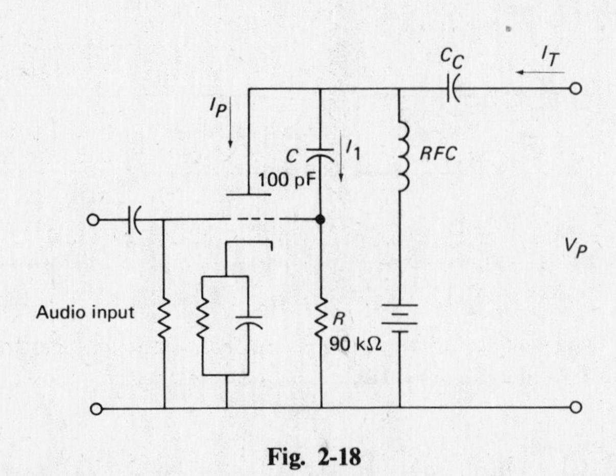

Fig. 2-18

2.60 Determine the swing in equivalent capacitance presented by the circuit of Fig. 2-18 if the tube used has a transconductance that ranges from 1500 μS to 1900 μS. *Ans.* 13.5 nF, 17.1 nF

2.61 If the resistance and capacitance values of Fig. 2-18 were changed to 110 kΩ and 150 pF and the range of transconductance of the tube was from 3100 μS to 4000 μS, calculate the range of equivalent capacitance presented by this circuit. *Ans.* 51.15 nF, 66 nF

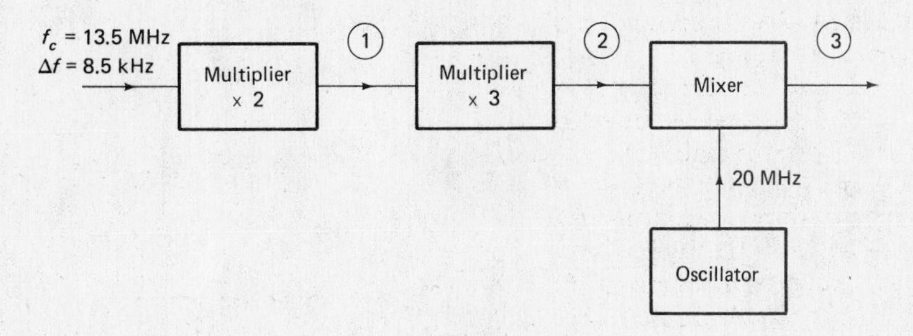

Fig. 2-19

2.62 Why can't crystal-controlled oscillators be used in the Direct method generation of frequency-modulated signals? *Ans.* The output frequency is not variable.

2.63 What is it about class C amplifiers and varactor diodes that permits their use as frequency multipliers?
Ans. They are nonlinear devices.

2.64 Figure 2-19 is the block diagram of the frequency-multiplication and heterodyne section of an FM transmitter. Determine the carrier frequency and frequency deviation at points 1, 2, and 3.
Ans. 27 MHz, 17 kHz, 81 MHz, 51 kHz; 101 MHz; 51 kHz

2.65 Repeat Problem 2.64 if the input carrier frequency is 12 MHz, the input frequency deviation is 6 kHz, and the oscillator frequency is 18 MHz. *Ans.* 24 MHz, 12 kHz; 72 MHz, 36 kHz; 90 MHz, 36 kHz

2.66 In Fig. 2-20, determine the appropriate multiplier values and oscillator frequency so as to provide an output FM signal having a carrier frequency of 106 MHz with a frequency deviation of 60 kHz if the input is an FM signal having a carrier frequency of 9 MHz and a frequency deviation of 10 kHz.
Ans. ×2, ×3, 52 MHz

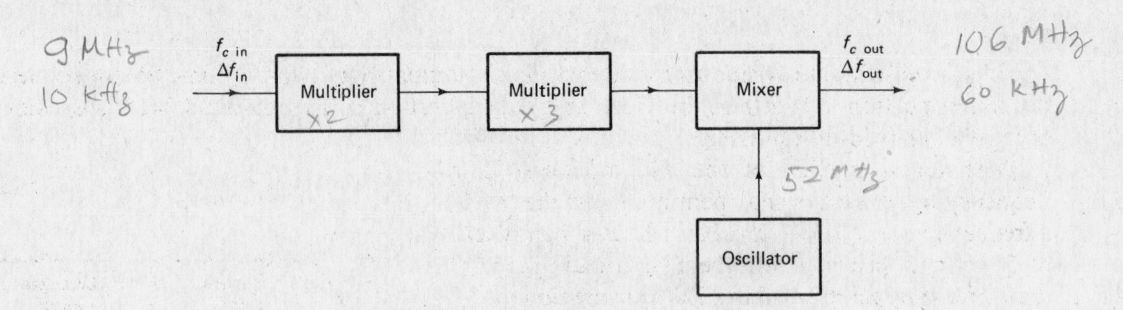

Fig. 2-20

2.67 The block diagram of Fig. 2-20 is required to produce an output FM signal having a carrier frequency of 54 MHz and a frequency deviation of 24 kHz when presented with an FM signal at its input having a carrier frequency of 7.0 MHz and a frequency deviation of 4 kHz. Determine the multiplier values and frequency the oscillator should be tuned to in order to meet these specifications.
Ans. ×3, ×2, 12 MHz

2.68 What is the difference between frequency multiplication and heterodyning?
Ans. When heterodyning, we are dealing with addition and subtraction.

Chapter 3

Television

Black-and-white television transmissions consist primarily of two portions, the video portion and the sound portion. Together, the video and sound portion of a commercial broadcast transmission occupy a bandwidth of 6 MHz.

The sound portion of the TV broadcast is a frequency-modulated signal permitted, by the FCC, a frequency deviation of 25 kHz. Thus a bandwidth of 50 kHz is provided for the FM sound signal at the extreme upper edge of the TV transmission.

The video portion of a television transmission contains synchronization information as well as information relevant to black-white variations.

That part of the television transmission relevant to the video portion of the signal is broadcast as a *vestigial-sideband signal.* A vestigial-sideband signal is basically an AM signal that has been doctored a bit. The information to be transmitted relevant to the video signal has a bandwidth of 4 MHz. If this were to be sent on a simple amplitude-modulated signal, an 8 MHz bandwidth would be required in order to accommodate the full upper and lower sidebands. Vestigial sideband was developed in order to reduce the required bandwidth and still permit transmission of the 4-MHz information content. See Fig. 3-1.

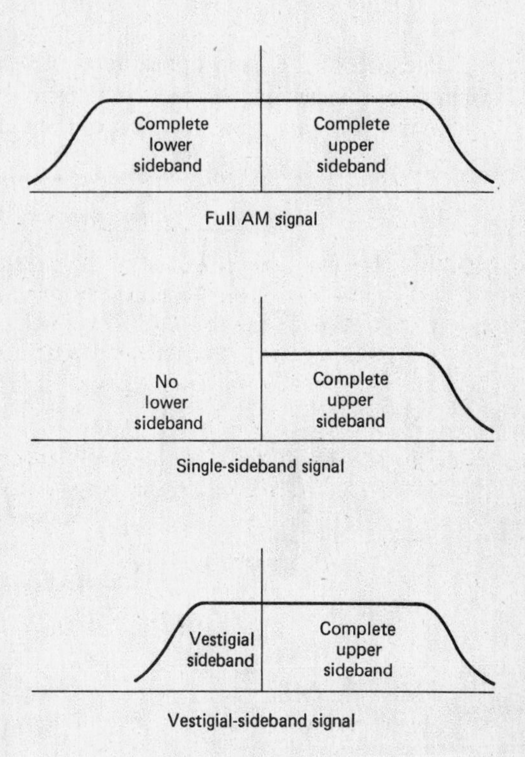

Fig. 3-1

One way of reducing the necessary bandwidth would have been simply to use single-sideband transmission, thereby requiring only a 4-MHz bandwidth. However, due to the problem at TV frequencies of being unable to sharply cut off one sideband without accidentally removing part of the sideband that is to remain, it is necessary to leave a portion of the removed sideband. When using vestigial-sideband transmission, one complete sideband, the carrier, and a portion of the other sideband is transmitted, thereby guaranteeing that the full information content is sent out. See Fig. 3-2.

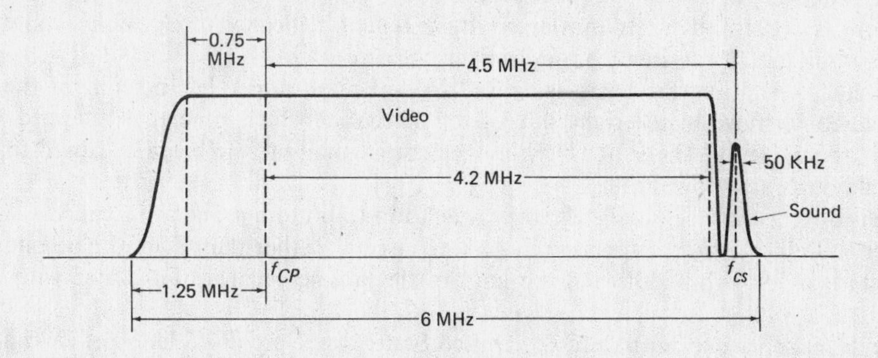

Fig. 3-2

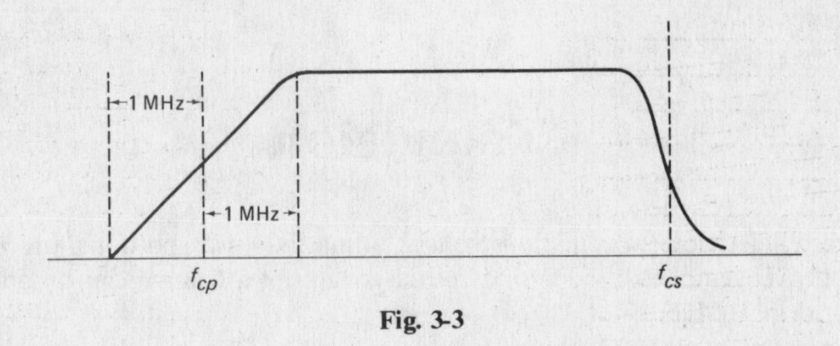

Fig. 3-3

Because of the dual appearance of *some* of the transmitted signal in the video signal in the vicinity of the carrier, the frequency response of the TV receiver is as shown in Fig. 3-3.

A time domain representation of the video portion of a TV signal is shown in Fig. 3-4.

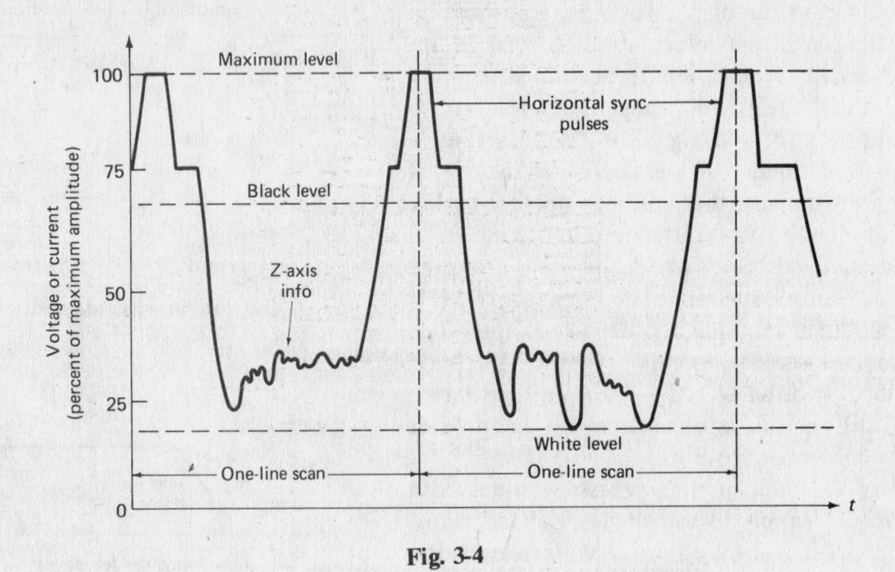

Fig. 3-4

HOW IT WORKS

An electron gun in the neck of the picture tube located in the TV receiver creates an electron beam which is accelerated down the tube, where it eventually strikes a luminescent screen that glows when struck by the electrons.

Black-to-white variations at each spot on the screen are obtained by controlling the voltage on a grid placed within the path of the electron beam.

The beam is controlled in its movement by magnetic deflection coils which cause it to move vertically and horizontally as seen from in front of the screen.

In the actual operation, the beam is caused to strike the upper left portion of the screen and then it is caused to move to the right side of the screen, at which time it is snapped back to the left side. All the while that the left-to-right movement is going on, the beam is caused to move down the screen, although at a slower rate.

Each left-to-right traverse is called a *line*. Each top-to-bottom sequence is called a *field*.

In order to reduce flicker, *interlaced scanning* is used. Rather than start the second field at the top left, it starts at the top middle and the lines of the second field are *interlaced* with those of the first.

Two fields make up one *frame*. In the United States, there are 262.5 lines per field and 525 lines per frame. See Fig. 3-5. There are 60 fields per second and thus 30 frames per second.

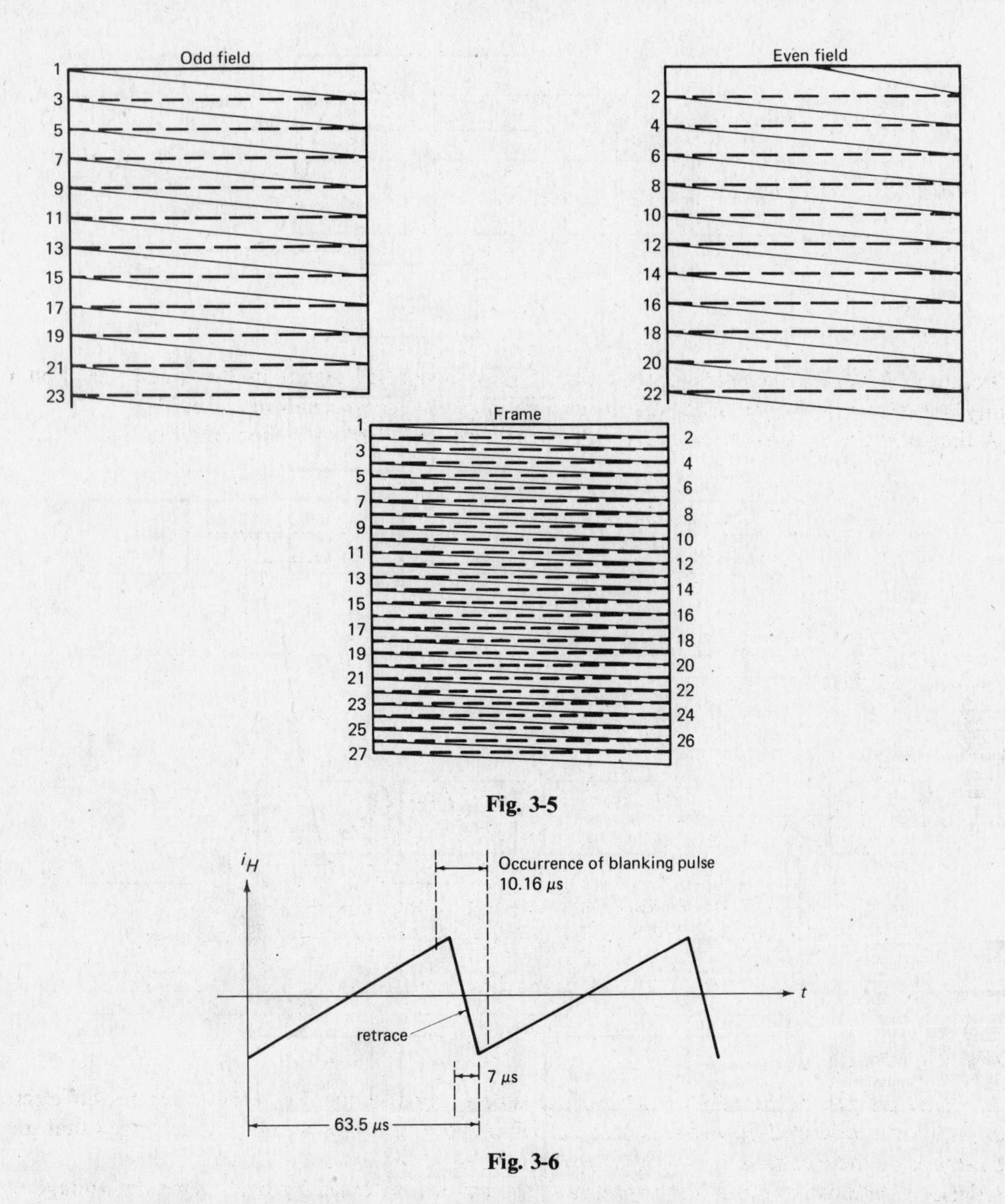

Fig. 3-5

Fig. 3-6

A sawtooth current waveshape with the geometry shown in Fig. 3-6 is passed through the horizontal deflection coil in order to obtain the appropriate timing.

Horizontal sync pulses are included within the picture portion of the TV transmission so as to synchronize the movements of the electron beam in the camera tube within the TV studio with the electron beam in the TV receiver.

A block diagram of a typical TV receiver is shown in Fig. 3-7, while the block diagram for a typical TV transmitter is shown as Fig. 3-8.

COLOR TELEVISION

The system of color television transmission and reception used in the United States is called the NTSC system. It was named for the National Television Systems Committee, which developed the system. The main requirement for the system in order for it to have been accepted was that it be

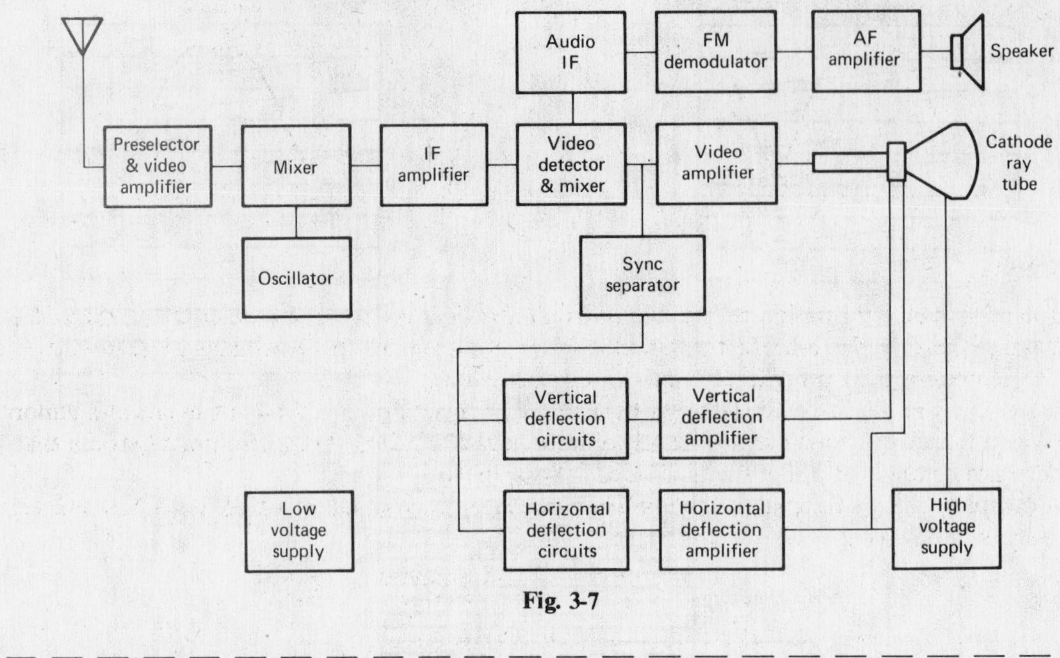

Fig. 3-7

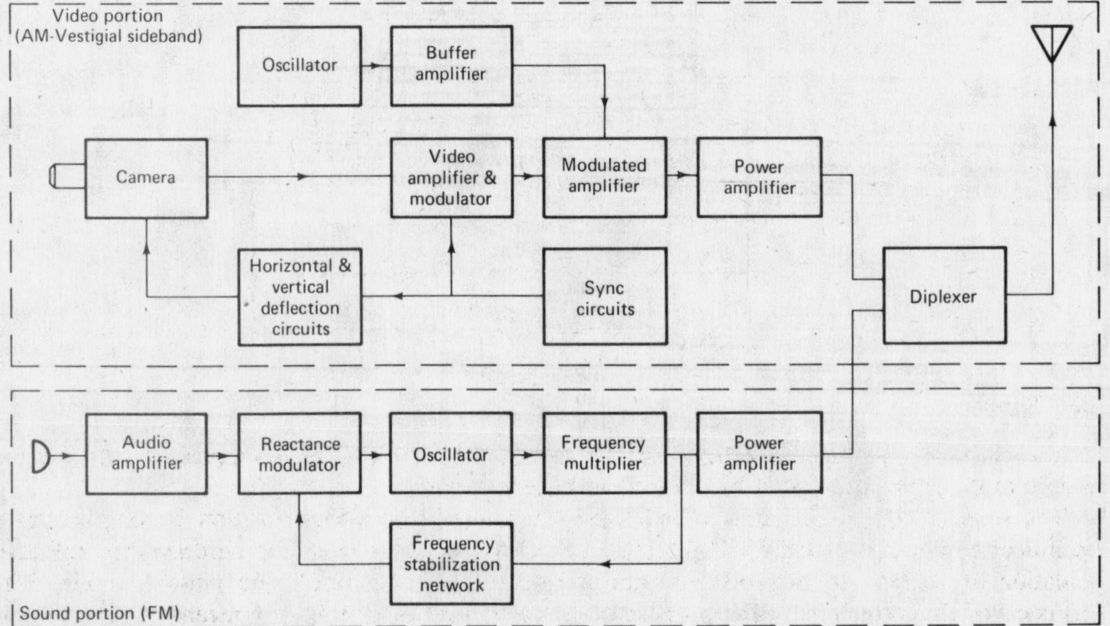

Fig. 3-8

compatible with the black-and-white system then (1953) in use and that the signal to be transmitted had to be within the 6-MHz-per-channel bandwidth allocation.

The color TV video transmission consists of two major portions, the *luminance* portion and the *chrominance* portion. The luminance portion of the video signal relates to the brightness of the scene being transmitted. Essentially this is what is contained in the video portion of the black-and-white TV transmission. The chrominance portion of the color TV signal contains information relevant to hue (color).

The TV receiver is informed of the color composition of each point contained in a scene by the transmission of information as to the amount of red, blue, and green contained at each point.

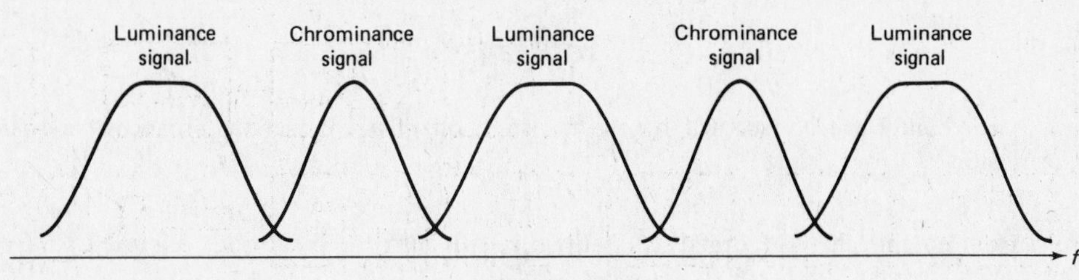

Fig. 3-9

Interleaving

The additional information relevant to hue that must be contained in a color transmission has to be contained within the 6-MHz bandwidth that is occupied by the black-and-white transmission. This is accomplished by using a technique called *interleaving.*

Interleaving makes use of the fact that the luminance signal does not fill up its bandwidth uniformly but instead there are spaces between bands of information. It is within these unused spaces that the chrominance component of the signal is fitted. See Fig. 3-9.

A complete color TV signal, including luminance signal, chrominance signal, and FM sound signal, is shown as Fig. 3-10.

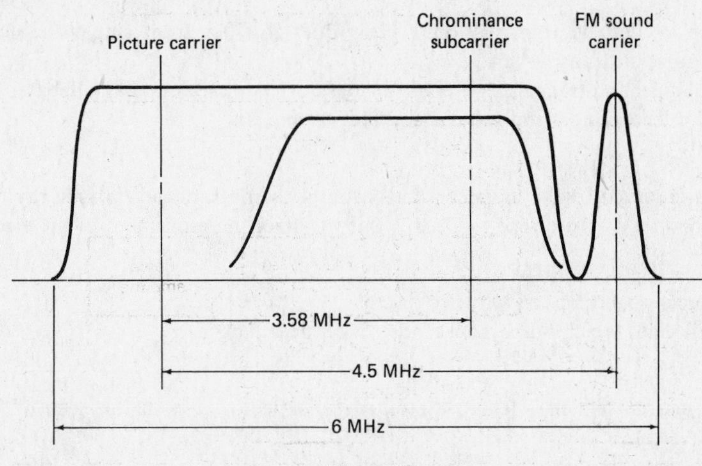

Fig. 3-10

Color Burst

The chrominance portion of the color TV signal is transmitted as a suppressed carrier signal. Because of this, the carrier must be reintroduced at the receiver.

The receiver contains what is called a *subcarrier oscillator*, whose function is to generate the chrominance subcarrier locally. Eight to eleven cycles of the transmitter chrominance subcarrier oscillation are caused to ride on the back portion of the horizontal sync pulse (see Fig. 3-11). This is called the *color burst*. The purpose of the color burst is to provide a means of synchronizing the subcarrier in the receiver with the one in the transmitter.

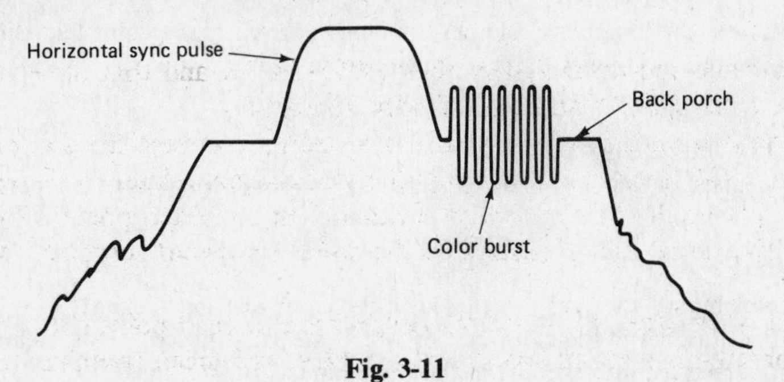

Fig. 3-11

Problems

3.1 How much bandwidth is occupied by the transmission of a commercial broadcast television station transmission? *Ans.* 6 MHz

3.2 How many commercial AM broadcast stations can fit into the bandwidth allocated to a commercial television channel? *Ans.* 600

3.3 How many commercial FM broadcast stations can fit into the bandwidth occupied by a commercial television station? *Ans.* 30

3.4 How much bandwidth is occupied by the sound portion of a TV transmission in the TV broadcast band? *Ans.* 50 kHz

3.5 What kind of modulation is used for the sound portion of a commercial broadcast TV transmission? *Ans.* FM

3.6 What is vestigial-sideband transmission? How does it differ from single-sideband transmission and amplitude-modulation transmission?
Ans. Vestigial-sideband transmission is a transmission in which one full sideband, the carrier, and only a portion of the second sideband are transmitted.

3.7 What function is served by the magnetic coils around the neck of the cathode ray tube which is used as the picture tube in a TV receiver? *Ans.* They control the movement of the electron beam.

3.8 What is interlaced scanning and why is it used?
Ans. Alternate lines which make up a field. It reduces flicker.

3.9 Why does the current which is fed to the magnetic deflector coils at the neck of a receiving tube have a sawtooth waveform?
Ans. So as to draw the beam across the screen at a uniform rate and then return it instantaneously.

3.10 What scheme is employed to cause the electron beam in the TV receiver and the electron beam in the studio camera to track identically? *Ans.* Synchronization pulses are transmitted.

3.11 What were the main requirements for a color television system to be acceptable?
Ans. It had to be compatible with preexisting black-and-white sets and require no more than a 6-MHz bandwidth.

3.12 Distinguish between the luminance and chrominance portion of the color television signal.
Ans. Luminance relates to brightness. Chrominance relates to hue.

3.13 What is meant by interleaving?
Ans. The placement of the chrominance signal in the band space between portions of the luminance signal.

3.14 What is the color burst? Why is it included in the color television signal?
Ans. A number of cycles of the color carrier which are transmitted. It is included to synchronize the subcarrier in the receiver with the one in the transmitter.

Chapter 4

Tuned Circuits

INTRODUCTION

There is a need to have frequency-selective circuits in communications electronics in order to be able to separate desired signals from undesired signals.

Series and parallel combinations of a capacitance and an inductance provide such frequency-selective circuits. See Fig. 4-1.

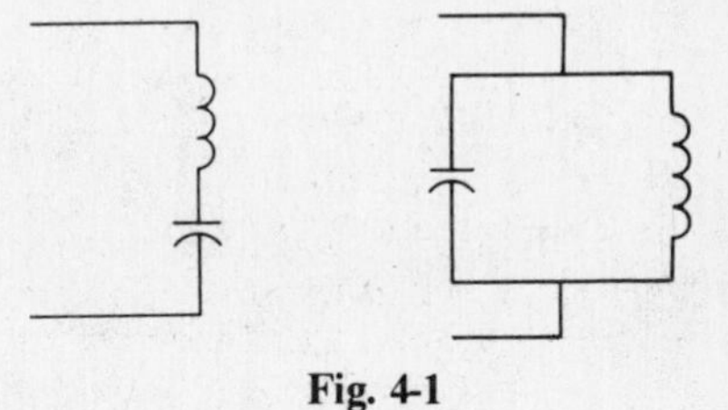

Fig. 4-1

THE SERIES LC CIRCUIT

A series combination of a capacitor and inductor with no resistance in the circuit has a frequency at which the impedance of the combination is zero.

This occurs when $X_L = X_C$.

The frequency of this occurrence, f_0, is called the resonant frequency of the circuit.

$$f_0 = \frac{1}{2\pi\sqrt{LC}}$$

THE SERIES RLC CIRCUIT

Since all circuits contain resistance, any meaningful study must examine the series combination of a capacitance, an inductance, and a resistance, the RLC circuit. See Fig. 4-2.

In the series RLC circuit, the minimum impedance reached, at the resonant frequency, is equal to the series resistance.

$$Z_0 = R$$

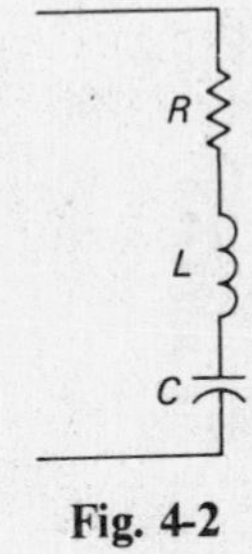

Fig. 4-2

The resonant frequency for the RLC circuit is the same as for the LC circuit:

$$f_0 = \frac{1}{2\pi\sqrt{LC}}$$

BANDWIDTH AND Q

A plot of impedance versus frequency for an RLC circuit results in a curve such as that shown in Fig. 4-3, while a plot of current versus frequency for an RLC circuit results in the curve of Fig. 4-4.

The exact shape of the curve for Z vs. f and I vs. f is dependent on the Q of the circuit, where Q is defined as

$$Q = \frac{X_L}{R}$$

for the simple *series* RLC circuit. See Fig. 4-5.

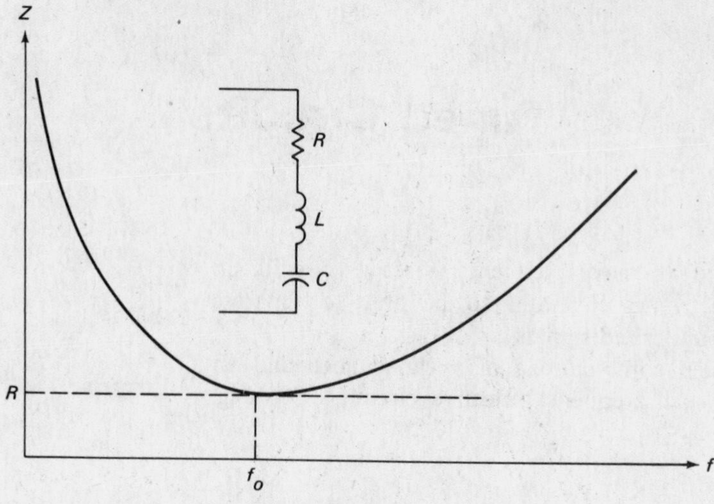

Fig. 4-3

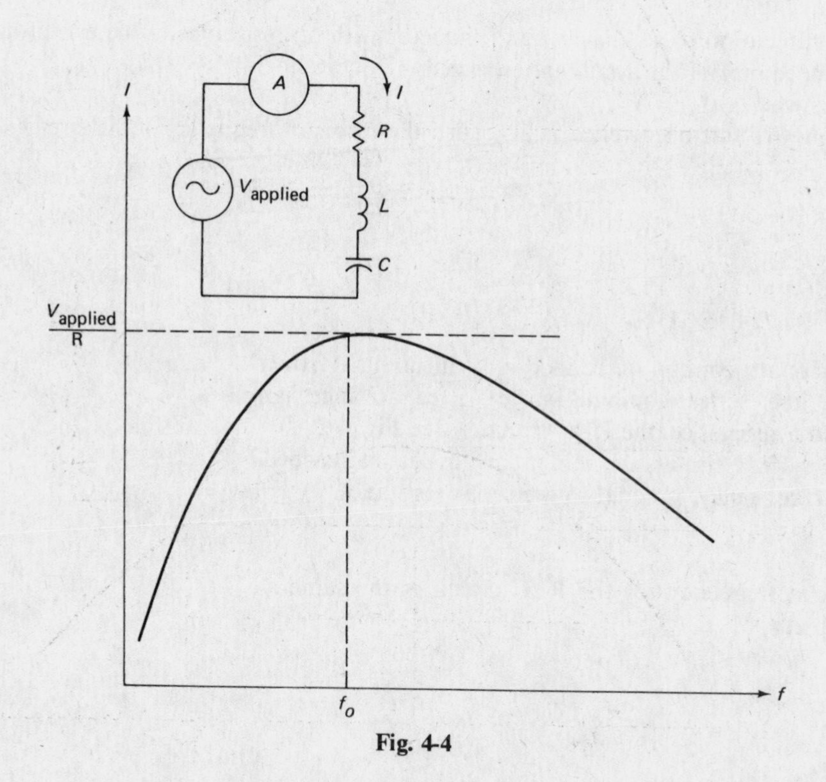

Fig. 4-4

The bandwidth of a series RLC circuit is defined as the difference between the two frequencies at which the current is down to 0.707 of the maximum current reached at resonance. These frequencies are referred to as the *upper* and *lower cutoff frequencies.* They are also called the *half power points.* See Fig. 4-6.

The bandwidth and Q of a circuit are related by

$$\mathrm{BW} = \frac{f_0}{Q}$$

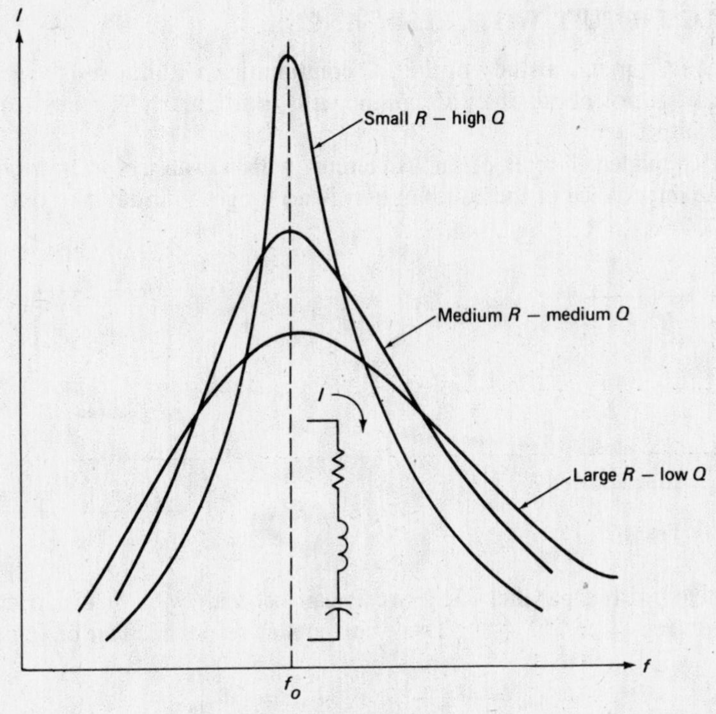

Fig. 4-5

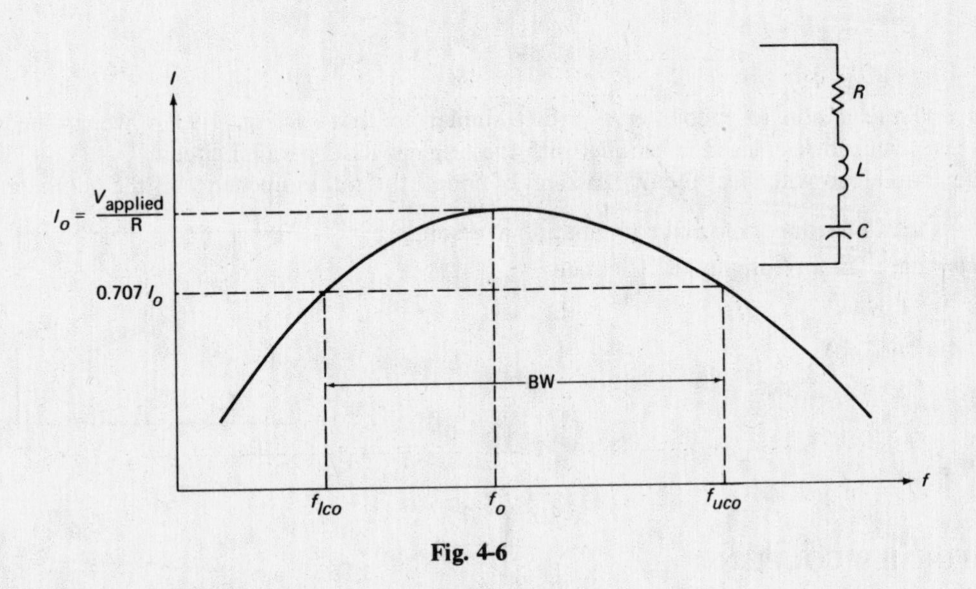

Fig. 4-6

THE PARALLEL LC CIRCUIT

An examination of the parallel combination of a capacitor and inductor indicates a resonance occurring at

$$f_0 = \frac{1}{2\pi\sqrt{LC}}$$

At this frequency, the impedance goes to infinity (assuming a resistanceless circuit).

THE PARALLEL LC CIRCUIT WITH RESISTANCE

As with the series LC circuit, a study of the LC combination without resistance is only of limited use since such a circuit is not obtainable; all components, particularly inductances have an inherent resistance associated with them.

Figure 4-7 is an equivalent circuit of an LC circuit with resistance. The resonant frequency is still $1/2\pi\sqrt{LC}$ but the impedance of the combination is no longer infinite at resonance but is equal to $Z_0 = QX_L$, where $Q = X_L/R$.

Fig. 4-7

Fig. 4-8

The current involved with a parallel RLC circuit can be dealt with in two parts, the line current and the circulating current. See Fig. 4-8. These currents at resonance can be found from

$$I_{0\ \text{line}} = \frac{V_{\text{applied}}}{Z_0} \qquad I_{\text{circ}} = QI_{0\ \text{line}}$$

The bandwidth of the circuit can be found from

$$\text{BW} = \frac{f_0}{Q}$$

It is not uncommon to encounter a circuit similar to that of Fig. 4-9, in which there is an external resistive load R_X placed in parallel with the original RLC parallel circuit.

The external load will affect the Q, BW, and of course the total impedance of the combination.

$Z_T \equiv$ total impedance of total combination at resonance

$Z_{10} =$ impedance of original RLC circuit

$$Z_T = \frac{Z_{10} R_X}{Z_{10} + R_X}$$

$$Q_T = \frac{Z_T}{X_L}$$

$$\text{BW} = \frac{f_0}{Q_T}$$

Fig. 4-9

TRANSFORMER COUPLING

A common coupling arrangement encountered in communications (RF) equipment is a coupling transformer. Figure 4-10 shows various tuned coupling arrangements in which a capacitor or capacitors are used in conjunction with a transformer to allow only the desired band of RF signals to pass. The coefficient of coupling of a transformer is a measure of how much of the magnetic flux originated by the primary links the secondary of the transformer. Although for *audio* equipment it is not unusual to find coefficients of coupling on the order of .90 and higher, transformers in RF service have coefficients of coupling on the order of .01 to .05.

RF transformer coupling arrangements have frequency response curves as shown in Figs. 4-11 and 4-12.

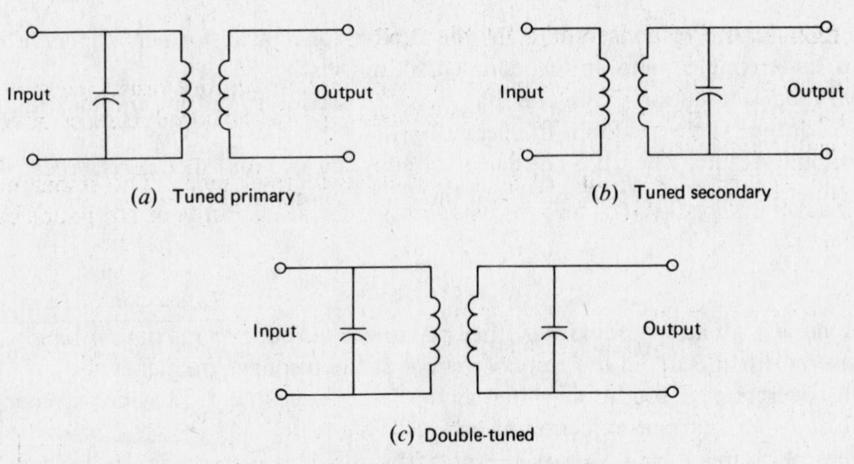

Fig. 4-10

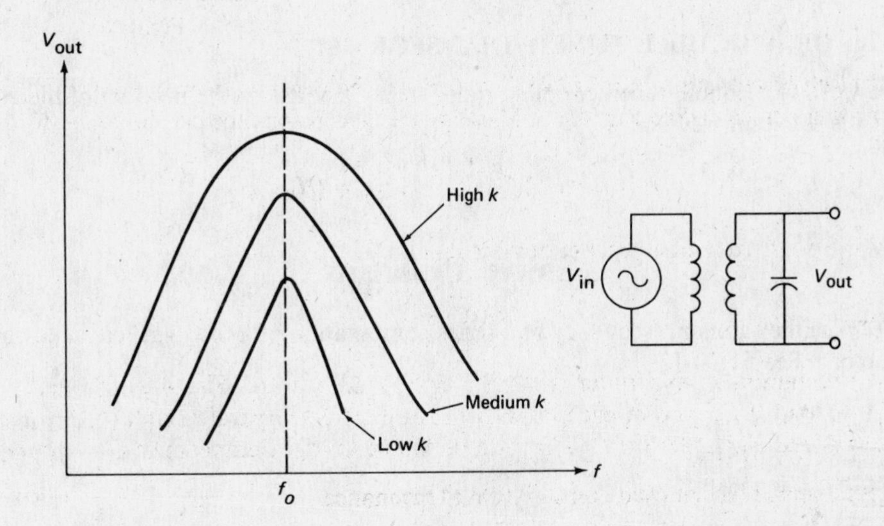

Fig. 4-11

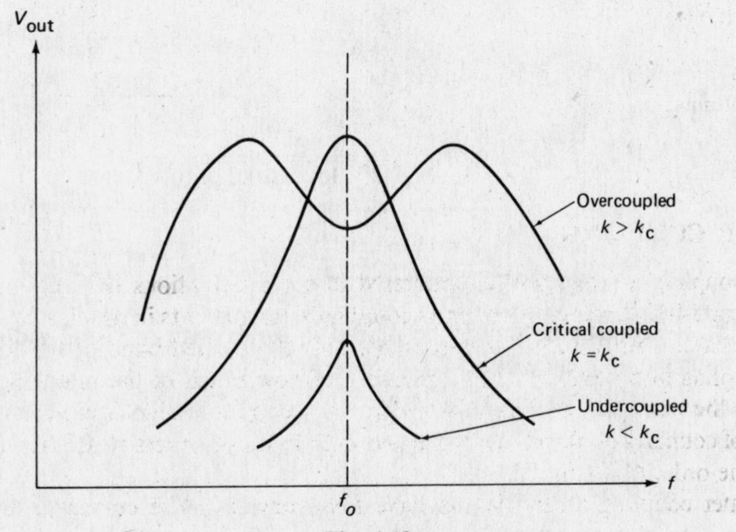

Fig. 4-12

Refer to Fig. 4-12, the response curve for the double-tuned transformer, and note the reference to overcoupled, undercoupled, and critical coupled conditions.

The *critical coupled* situation is the one that provides maximum output and maximum bandwidth without a dip in output at the resonant frequency.

Critical coupling results when the coefficient of coupling is equal to the reciprocal of the square root of the product of the Q of the primary and the Q of the secondary.

$$k_c = \frac{1}{\sqrt{Q_P Q_S}}$$

The undercoupled situation provides neither maximum output nor maximum bandwidth. In the overcoupled case, a dip appears in the response curve at the resonant frequency.

The slightly overcoupled condition is frequently desired because it provides steeper sides to the response curve and thus sharpens rejection of undesired signals.

A coefficient of coupling of 1.5 times the critical coefficient of coupling is frequently used and considered desirable.

BANDWIDTH OF A DOUBLE-TUNED TRANSFORMER

The bandwidth of a double-tuned coupling transformer is equal to the product of the coefficient of coupling and the resonant frequency.

$$\text{BW} = kf_0$$

Solved Problems

4.1 Determine the resonant frequency of a series combination of a 0.001-μF capacitor and a 2-mH inductor. See Fig. 4-13.

SOLUTION

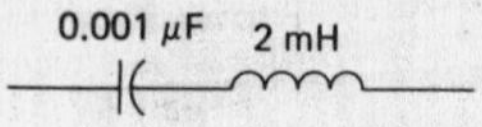

Given: $C = 0.001\ \mu\text{F}$

$L = 2\ \text{mH}$

Find: f_0

0.001 μF 2 mH

Fig. 4-13

Using the formula for resonant frequency,

$$f_0 = \frac{1}{2\pi\sqrt{LC}}$$

and substituting,

$$f_0 = \frac{1}{2\pi\sqrt{(2 \times 10^{-3})(0.001 \times 10^{-6})}}$$

$$\boxed{f_0 = 112.6\ \text{kHz}}$$

4.2 What value of capacitance should be placed in series with a 27-μH inductance in order to obtain a resonant frequency of 5 MHz? See Fig. 4-14.

SOLUTION

Given: $L = 27\ \mu\text{H}$

$f_0 = 5\ \text{MHz}$

Find: C

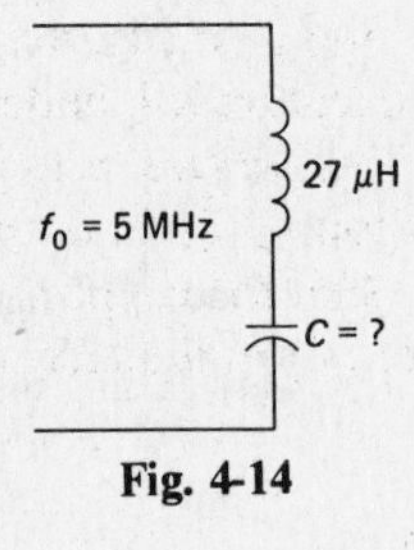

Fig. 4-14

The formula for resonant frequency is

$$f_0 = \frac{1}{2\pi\sqrt{LC}}$$

Solving for C, square both sides of the equation:

$$f_0^2 = \frac{1}{4\pi^2 LC}$$

Then multiply both sides of the equation by C/f_0^2:

$$C = \frac{1}{4\pi^2 f_0^2 L}$$

Substituting in this equation,

$$C = \frac{1}{4\pi^2(5 \times 10^6)^2(27 \times 10^{-6})}$$

$$= \frac{1}{4\pi^2(25 \times 10^{12})(27 \times 10^{-6})}$$

$$= 37.56 \times 10^{-12}$$

$$\boxed{C = 37.56 \text{ pF}}$$

4.3 A series circuit containing 150 pF of capacitance is to be made resonant at 75 MHz. Determine the inductance which must be placed in series with the capacitance. See Fig. 4-15.

SOLUTION

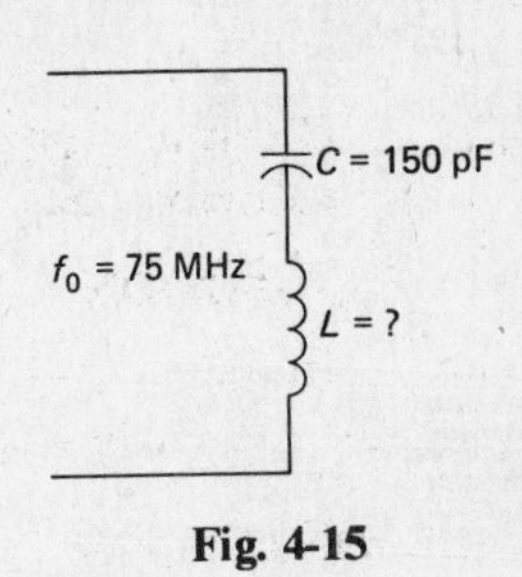

Fig. 4-15

Given: $C = 150$ pF
$f_0 = 75$ MHz

Find: L

Begin with the formula for resonant frequency:

$$f_0 = \frac{1}{2\pi\sqrt{LC}}$$

Solve for L by first squaring both sides of the equation and transposing appropriate quantities:

$$L = \frac{1}{4\pi^2 f_0^2 C}$$

$$= \frac{1}{4\pi^2(75 \times 10^6)^2(150 \times 10^{-12})}$$

$$= 0.03 \times 10^{-6}$$

$$\boxed{L = 0.03\ \mu\text{H}}$$

4.4 (*a*) Determine the resonant frequency of a circuit consisting of a simple series combination of a 0.001-μF capacitor and a 16-μH inductor. See Fig. 4-16.
(*b*) What is the impedance of the circuit at this resonant frequency assuming no resistance.
(*c*) Calculate the reactance of each of the components of this circuit at resonance.

Fig. 4-16

SOLUTION

Given: $C = 0.001\ \mu F$
$L = 16\ \mu H$

Find: (a) f_0 (b) Z_0 (c) X_{LO}, X_{CO}

(a) Using the equation for resonant frequency,

$$f_0 = \frac{1}{2\pi\sqrt{LC}}$$
$$= \frac{1}{2\pi\sqrt{(16 \times 10^{-6})(0.001 \times 10^{-6})}}$$
$$= 1.259 \times 10^6$$

$$\boxed{f_0 = 1.259 \text{ MHz}}$$

(b) The impedance of a resistanceless series combination of a series LC circuit at resonance is zero. Therefore

$$\boxed{Z_0 = 0}$$

(c) The formula for inductive reactance is

$$X_L = 2\pi f L$$

Substituting,

$$X_{LO} = 2\pi(1.259 \times 10^6)(16 \times 10^{-6})$$

$$\boxed{X_{LO} = 126.5\ \Omega}$$

Calculating capacitive reactance,

$$X_C = \frac{1}{2\pi f C}$$

$$X_{CO} = \frac{1}{2\pi(1.259 \times 10^6)(0.001 \times 10^{-6})}$$

$$\boxed{X_{CO} = 126.5\ \Omega}$$

4.5 A series circuit consists of a 15-Ω resistance, a 0.01-mH inductance, and a 0.01-μF capacitance. See Fig. 4-17.

(a) Calculate the resonant frequency of the circuit.
(b) What is the impedance of the combination at the resonant frequency?
(c) How much current would flow if a 10-V source tuned to the resonant frequency were applied to this circuit?

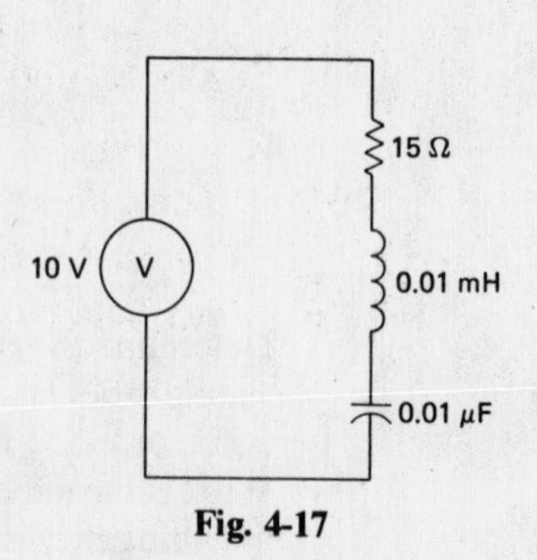

Fig. 4-17

SOLUTION

Given: $R = 15\ \Omega$
$L = 0.01$ mH
$C = 0.01\ \mu F$
$V_0 = 10$ V

Find: (a) f_0 (b) Z_0 (c) I_0

(*a*) Use the resonant-frequency formula,

$$f_0 = \frac{1}{2\pi\sqrt{LC}}$$
$$= \frac{1}{2\pi\sqrt{(0.01 \times 10^{-3})(0.01 \times 10^{-6})}}$$
$$= 503.55 \times 10^3$$

$$\boxed{f_0 = 503.55 \text{ kHz}}$$

(*b*) At resonance, the total impedance of a series RLC circuit is equal to the resistance.

$$Z_0 = R$$

$$\boxed{Z_0 = 15\ \Omega}$$

(*c*) By Ohm's law,

$$I_0 = \frac{V_0}{Z_0}$$
$$= \frac{10}{15}$$
$$= 0.667 \text{ A}$$

$$\boxed{I_0 = 667 \text{ mA}}$$

4.6 Determine the following for the circuit shown as Fig. 4-18:

(*a*) The resonant frequency
(*b*) The total impedance at resonance
(*c*) The current that flows at resonance
(*d*) The inductive reactance at resonance
(*e*) The capacitive reactance at resonance
(*f*) The voltage across the resistor at resonance
(*g*) The voltage across the inductor at resonance
(*h*) The voltage across the capacitor at resonance

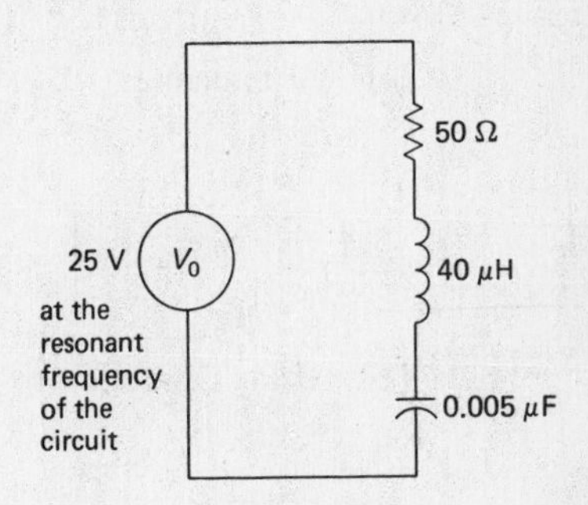

Fig. 4-18

SOLUTION

Given: $R = 50\ \Omega$
$L = 40\ \mu\text{H}$
$C = 0.005\ \mu\text{F}$
$V_0 = 25\text{ V}$

Find: (*a*) f_0 (*b*) Z_0 (*c*) I_0 (*d*) X_{L0} (*e*) X_{C0} (*f*) V_{R0} (*g*) V_{L0} (*h*) V_{C0}

(*a*) Using the equation for resonant frequency,

$$f_0 = \frac{1}{2\pi\sqrt{LC}}$$
$$= \frac{1}{2\pi\sqrt{(40 \times 10^{-6})(0.005 \times 10^{-6})}}$$
$$= 3.56 \times 10^5$$
$$= 356 \times 10^3$$

$$\boxed{f_0 = 356 \text{ kHz}}$$

(*b*) At resonance, the total impedance is equal to the resistance. So,

$$Z_0 = R$$

$$\boxed{Z_0 = 50 \ \Omega}$$

(*c*) Using Ohm's law,

$$I_0 = \frac{V_0}{Z_0} = \frac{25}{50} = 0.5 \text{ A}$$

$$\boxed{I_0 = 500 \text{ mA}}$$

(*d*) The inductive reactance is

$$X_L = 2\pi f L$$

$$X_{L0} = 2\pi(356 \times 10^3)(40 \times 10^{-6})$$

$$\boxed{X_{L0} = 89.43 \ \Omega}$$

(*e*) At resonance, $X_L = X_C$,

$$X_{C0} = X_{L0} = 89.43$$

$$\boxed{X_{C0} = 89.43 \ \Omega}$$

(*f*) Using Ohm's law to determine the voltage across the resistance at resonance,

$$V_{R0} = I_0 R = 0.5(50)$$

$$\boxed{V_{R0} = 25 \text{ V}}$$

(*g*) Using Ohm's law again,

$$V_{L0} = I_0 X_{L0} = 0.5(89.43)$$

$$\boxed{V_{L0} = 44.72 \text{ V}}$$

(*h*) Once more, using Ohm's law,

$$V_{C0} = I_0 X_{C0} = 0.5(89.43)$$

$$\boxed{V_{C0} = 44.72 \text{ V}}$$

4.7 (a) Find the appropriate value of capacitance required in order to provide a series resonant frequency of 90 MHz when using a 40-μH inductance which has a resistance of 80 Ω. See Fig. 4-19.
(b) What is the Q of the circuit?
(c) Calculate the bandwidth of the circuit.

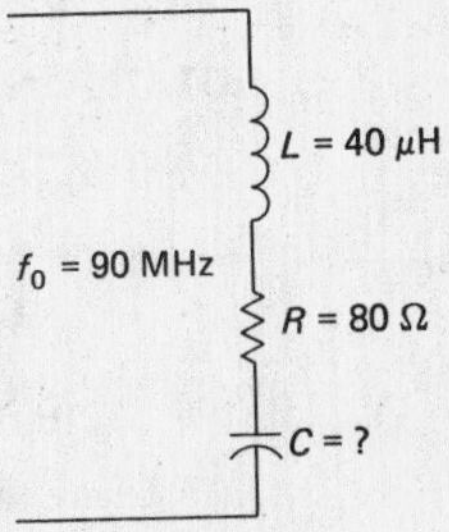

Fig. 4-19

SOLUTION

Given: $f_0 = 90$ MHz
$L = 40\ \mu H$
$R = 80\ \Omega$

Find: (a) C (b) Q (c) BW

(a) The equation for determining resonant frequency is

$$f_0 = \frac{1}{2\pi\sqrt{LC}}$$

Squaring both sides of the equation,

$$f_0^2 = \frac{1}{4\pi^2 LC}$$

Solving for C,

$$C = \frac{1}{4\pi^2 f_0^2 L}$$

Substituting,

$$C = \frac{1}{4\pi^2(90 \times 10^6)^2(40 \times 10^{-6})}$$
$$= 7.83 \times 10^{-14}$$

$$\boxed{C = 0.0783 \text{ pF}}$$

(b) By definition,

$$Q = \frac{X_L}{R}$$

Replacing X_L with $2\pi f L$,

$$Q = \frac{2\pi f_0 L}{R}$$

Substituting,

$$Q = \frac{2\pi(90 \times 10^6)(40 \times 10^{-6})}{80}$$

$$\boxed{Q = 282.6}$$

(c)

$$\text{BW} = \frac{f_0}{Q}$$
$$= \frac{90 \times 10^6}{282.6}$$
$$= 318.5 \times 10^3$$

$$\boxed{\text{BW} = 318.5 \text{ kHz}}$$

4.8 (*a*) A series circuit is desired which will be resonant at 150 MHz. A capacitance of 50 pF exists in the circuit. See Fig. 4-20. Determine the value of inductance required.
(*b*) The inductor used is to be chosen from a group of inductances all of which have a Q of 50. What is the bandwidth of this circuit?

SOLUTION

Given: $f_0 = 150$ MHz
$C = 50$ pF
$Q = 50$

Find: (*a*) L (*b*) BW

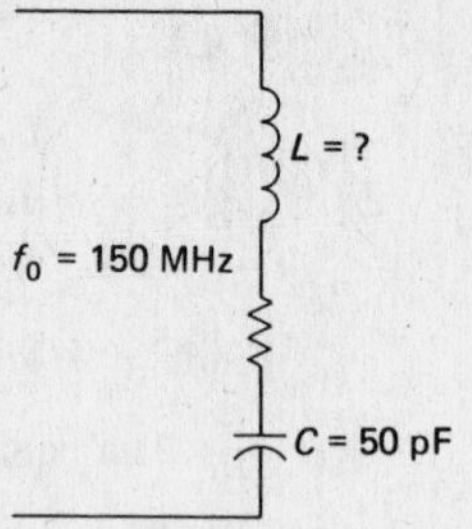

L = ?
f_0 = 150 MHz
C = 50 pF

Fig. 4-20

(*a*) Using the basic resonant-frequency equation and solving for L,

$$f_0 = \frac{1}{2\pi\sqrt{LC}}$$

$$f_0^2 = \frac{1}{4\pi^2 LC}$$

$$L = \frac{1}{4\pi^2 f_0^2 C}$$

$$= \frac{1}{4\pi^2(150 \times 10^6)^2(50 \times 10^{-12})}$$

$$\boxed{L = 0.225\ \mu\text{H}}$$

(*b*)

$$\text{BW} = \frac{f_0}{Q}$$

$$= \frac{150 \times 10^6}{50}$$

$$\boxed{\text{BW} = 3\text{ MHz}}$$

4.9 Calculate the resistance of a series RLC circuit which is resonant at 50 MHz and has a Q of 100. The inductance of the circuit is 150 μH. See Fig. 4-21.

SOLUTION

Given: $f_0 = 50 \times 10^6$ Hz
$Q = 100$
$L = 150\ \mu$H

Find: R

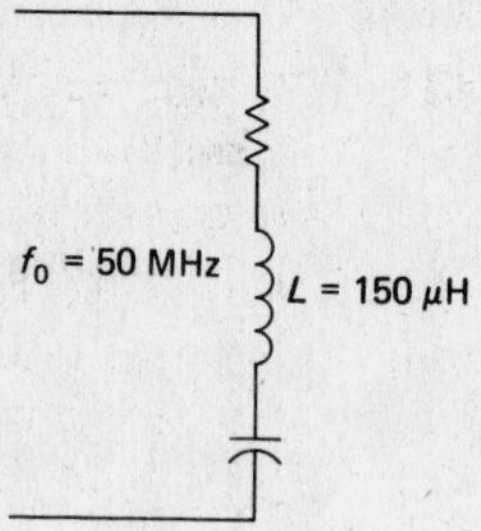

f_0 = 50 MHz
L = 150 μH

Fig. 4-21

Since

$$Q = \frac{X_L}{R}$$

$$R = \frac{X_L}{Q}$$

Substituting $2\pi f_0 L$ for X_L,

$$R = \frac{2\pi f_0 L}{Q}$$

$$= \frac{2\pi(50 \times 10^6)(150 \times 10^{-6})}{100}$$

$$\boxed{R = 471\ \Omega}$$

4.10 Find the inductance in a series circuit which is resonant to 250 MHz and which has a bandwidth of 10 MHz. The circuit has a resistance of 50 Ω. See Fig. 4-22.

SOLUTION

Given: $f_0 = 250$ MHz
BW = 10 MHz
$R = 50\ \Omega$

Find: L

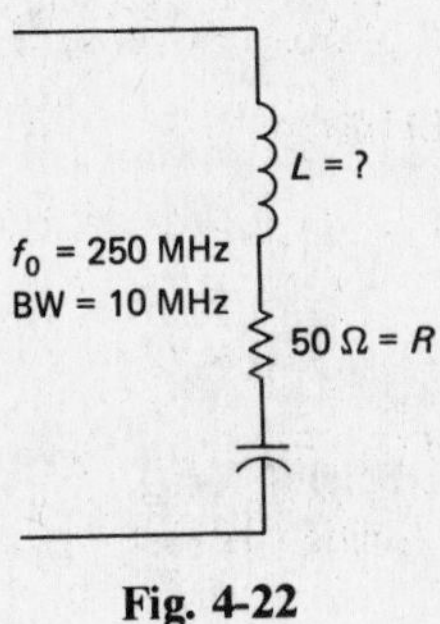

Fig. 4-22

Using the formula for bandwidth, $\text{BW} = f_0/Q$,

$$10 \times 10^6 = \frac{250 \times 10^6}{Q}$$

$$Q = \frac{250 \times 10^6}{10 \times 10^6} = 25$$

The formula relating Q, X_L, and R for a series circuit can now be used to find L.

$$Q = \frac{X_L}{R}$$

Substituting $2\pi f_0 L$ for X_L,

$$Q = \frac{2\pi f_0 L}{R}$$

Solving for L,

$$L = \frac{RQ}{2\pi f_0} = \frac{50(25)}{2(3.14)(250 \times 10^6)} = 0.796 \times 10^{-6}$$

$$\boxed{L = 0.796\ \mu\text{H}}$$

4.11 A parallel resonant circuit consists of an inductor of 50 μH in parallel with a 0.002 μF capacitor. The inductor has an internal resistance of 5 Ω. A 500-mV signal at the resonant frequency is applied across this circuit. See Fig. 4-23.

(*a*) Determine the frequency to which this circuit is resonant.
(*b*) Calculate the Q of the circuit.
(*c*) How much impedance does this circuit present at resonance?
(*d*) Find the line current when the 500-mV signal is applied.
(*e*) How big is the circulating current?
(*f*) What is the bandwidth of this circuit?

SOLUTION

Given: $L = 50\ \mu\text{H}$
$C = 0.002\ \mu\text{F}$
$R = 5\ \Omega$
$V_0 = 500$ mV

Find: (*a*) f_0 (*c*) Z_0 (*e*) I_{circ}
(*b*) Q (*d*) I_{L0} (*f*) BW

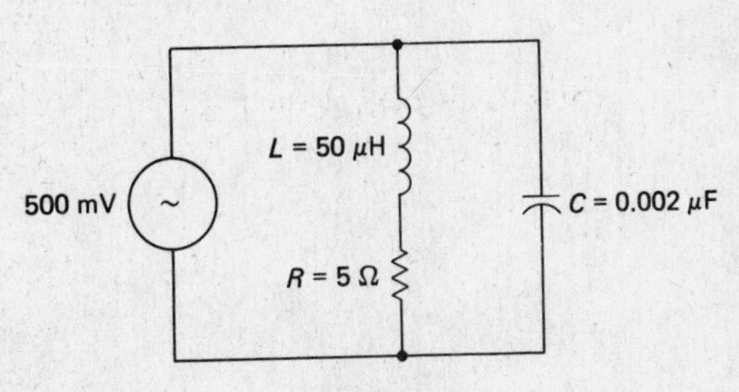

Fig. 4-23

(*a*) Using the resonant-frequency equation,

$$f_0 = \frac{1}{2\pi\sqrt{LC}}$$

$$= \frac{1}{2\pi\sqrt{(50 \times 10^{-6})(0.002 \times 10^{-6})}}$$

$$\boxed{f_0 = 503.5 \text{ kHz}}$$

(*b*) Solving for Q,

$$Q = \frac{X_L}{R}$$

$$= \frac{2\pi f_0 L}{R}$$

$$= \frac{2\pi(503.5 \times 10^3)(50 \times 10^{-6})}{5}$$

$$\boxed{Q = 31.62}$$

(*c*)

$$Z_0 = QX_L$$

$$= Q(2\pi f_0 L)$$

$$= (31.62)(2)(3.14)(503.5 \times 10^3)(50 \times 10^{-6})$$

$$\boxed{Z_0 = 5000 \ \Omega}$$

(*d*) Solving for line current using Ohm's law and the resonant-frequency impedance of the parallel combination,

$$I_{LO} = \frac{V_0}{Z_0}$$

$$= \frac{500 \times 10^{-3}}{5000}$$

$$\boxed{I_{LO} = 100 \ \mu\text{A}}$$

(*e*)

$$I_{\text{circ}} = QI_{LO}$$

$$= 31.62(100 \times 10^{-6})$$

$$\boxed{I_{\text{circ}} = 3.162 \text{ mA}}$$

(*f*)

$$\text{BW} = \frac{f_0}{Q}$$

$$= \frac{503.5 \times 10^3}{31.62}$$

$$\boxed{\text{BW} = 15.92 \text{ kHz}}$$

4.12 A 400-pF capacitor is in parallel with a coil which has an inductance of 200 μH and a resistance of 15 Ω. See Fig. 4-24.

(*a*) What is the resonant frequency of this circuit?
(*b*) Find the Q of the circuit.
(*c*) What is the impedance of this circuit at resonance?
(*d*) Determine the line current when a 2-V source is applied. The frequency of the source is the same as the resonant frequency of the circuit.
(*e*) Calculate the circulating current for the condition of 2 V applied as in (*d*).
(*f*) Find the bandwidth.

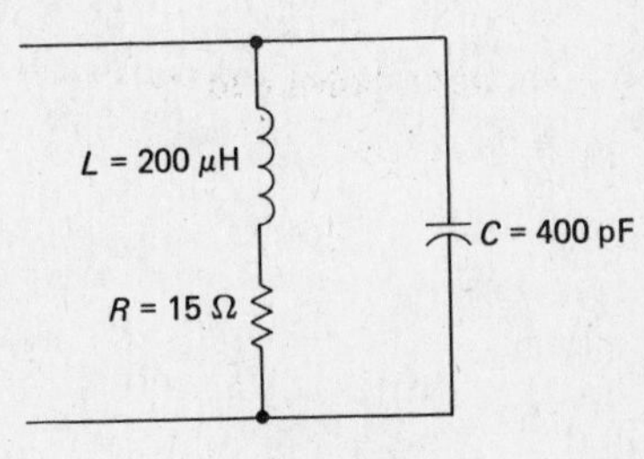

Fig. 4-24

SOLUTION

Given: $C = 400$ pF
$L = 200\ \mu$H
$R = 15\ \Omega$
$V_0 = 2$ V

Find: (*a*) f_0 (*c*) Z_0 (*e*) I_{circ}
(*b*) Q (*d*) I_{LO} (*f*) BW

(*a*) Using the resonant frequency formula,

$$f_0 = \frac{1}{2\pi\sqrt{LC}}$$
$$= \frac{1}{2\pi\sqrt{(200 \times 10^{-6})(400 \times 10^{-12})}}$$
$$= 563 \times 10^3$$

$$\boxed{f_0 = 563 \text{ kHz}}$$

(*b*)

$$Q = \frac{X_L}{R}$$
$$= \frac{2\pi f_0 L}{R}$$
$$= \frac{2\pi(563 \times 10^3)(200 \times 10^{-6})}{15}$$

$$\boxed{Q = 47.14}$$

(*c*)

$$Z_0 = QX_L$$
$$= 47.14(2\pi f_0 L)$$
$$= 47.14(2\pi)(563 \times 10^3)(200 \times 10^{-6})$$
$$= 33.33 \times 10^3$$

$$\boxed{Z_0 = 33.33 \text{ k}\Omega}$$

(*d*)

$$I_{LO} = \frac{V_0}{Z_0}$$
$$= \frac{2}{33.33 \times 10^3}$$
$$= 60 \times 10^{-6}$$

$$\boxed{I_{LO} = 60\ \mu\text{A}}$$

(e)
$$I_{\text{circ}} = QI_{LO}$$
$$= 47.14(60 \times 10^{-6})$$
$$= 2.83 \times 10^{-3}$$

$$\boxed{I_{\text{circ}} = 2.83 \text{ mA}}$$

(f)
$$\text{BW} = \frac{f_0}{Q}$$
$$= \frac{563 \times 10^3}{47.14}$$
$$= 11.94 \times 10^3$$

$$\boxed{\text{BW} = 11.94 \text{ kHz}}$$

4.13 An external resistive load R_X is installed in parallel with an RLC circuit as shown in Fig. 4-25.

(a) Find the frequency at which the circuit is resonant.
(b) What was the Q of the circuit prior to installation of the external load?
(c) Find the impedance of the circuit at resonance prior to installing the external resistance.
(d) Determine the impedance of the total circuit at resonance after the external resistance is installed.
(e) Find the Q of the circuit after installing the external resistance.
(f) Calculate the bandwidth of the circuit prior to installation of the external resistance.
(g) What is the bandwidth of the circuit after installing the external load?

SOLUTION

Given:
$R = 10\ \Omega$
$L = 0.05$ mH
$C = 150$ pF
$R_X = 20$ kΩ

Find:
(a) f_0 (e) Q_T
(b) Q_1 (f) BW_1
(c) Z_{10} (g) BW_2
(d) Z_T

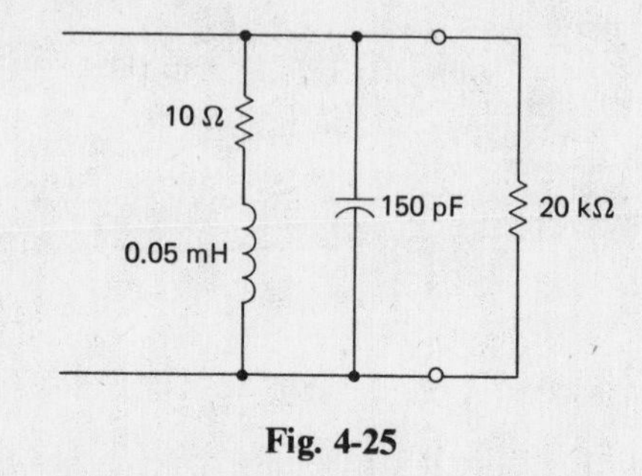

Fig. 4-25

(a) The resonant frequency is found by using the standard resonant-frequency formula,

$$f_0 = \frac{1}{2\pi\sqrt{LC}}$$
$$= \frac{1}{2\pi\sqrt{(0.05 \times 10^{-3})(150 \times 10^{-12})}}$$
$$= 1.839 \times 10^6$$

$$\boxed{f_0 = 1.839 \text{ MHz}}$$

(b) To determine the Q of the circuit prior to installing R_X, use the basic defining equation for Q,

$$Q = \frac{X_L}{R}$$

$$Q_1 = \frac{2\pi f_0 L}{R}$$

$$= \frac{2\pi(1.839 \times 10^6)(0.05 \times 10^{-3})}{10}$$

$$\boxed{Q_1 = 57.74}$$

(*c*)

$$Z_{10} = Q_1 X_L$$

$$= (57.74)[(2\pi)(1.839 \times 10^6)(0.05 \times 10^{-3})]$$

$$= 33.34 \times 10^3$$

$$\boxed{Z_{10} = 33.34\ \text{k}\Omega}$$

(*d*) In order to find the total impedance presented by the total circuit with the external resistance in the circuit, consider the circuit without the external resistance to be in parallel with the external resistance. Thus

$$Z_T = \frac{Z_{10} R_X}{Z_{10} + R_X}$$

$$= \frac{(33.34 \times 10^3)(20 \times 10^3)}{(33.34 \times 10^3) + (20 \times 10^3)}$$

$$= 12.5 \times 10^3$$

$$\boxed{Z_T = 12.5\ \text{k}\Omega}$$

(*e*) In order to find the Q of the circuit after installation of the external resistance, use the formula for Q_T.

$$Q_T = \frac{Z_T}{X_L}$$

$$= \frac{Z_T}{2\pi f_0 L}$$

$$= \frac{12.5 \times 10^3}{2\pi(1.839 \times 10^6)(0.05 \times 10^{-3})}$$

$$\boxed{Q_T = 21.65}$$

(*f*) To find the bandwidth prior to installing the external resistance, use the bandwidth formula with the Q obtained for the condition without the external resistance in (*b*).

$$\text{BW}_1 = \frac{f_0}{Q_1}$$

$$= \frac{1.839 \times 10^6}{57.74}$$

$$= 31.85 \times 10^3$$

$$\boxed{\text{BW}_1 = 31.85\ \text{kHz}}$$

(*g*) Use the bandwidth formula with the Q found for the total circuit in (*e*) to determine the bandwidth of the total circuit.

$$\begin{aligned} BW_2 &= \frac{f_0}{Q_T} \\ &= \frac{1.839 \times 10^6}{21.65} \\ &= 84.94 \times 10^3 \end{aligned}$$

$$\boxed{BW_2 = 84.94\ \text{kHz}}$$

4.14 A parallel RLC circuit as shown in Fig. 4-26 is to have a 6-kΩ resistive load applied as shown.

(*a*) Determine the resonant frequency of the circuit.
(*b*) Calculate the Q of the circuit prior to connecting the external load.
(*c*) What is the impedance at resonance before the external load is applied?
(*d*) Find the impedance of the total circuit at resonance after the external resistance is attached.
(*e*) Determine the Q of the circuit after installing the external resistance.
(*f*) Find the bandwidth of the circuit before the external resistive load is connected.
(*g*) How large is the bandwidth of the circuit after the external load is connected?

SOLUTION

Given: $R = 2.5\ \Omega$
$L = 30\ \mu\text{H}$
$C = 175\ \text{pF}$
$R_X = 6\ \text{k}\Omega$

Find: (*a*) f_0 (*e*) Q_T
(*b*) Q_{10} (*f*) BW_1
(*c*) Z_{10} (*g*) BW_2
(*d*) Z_T

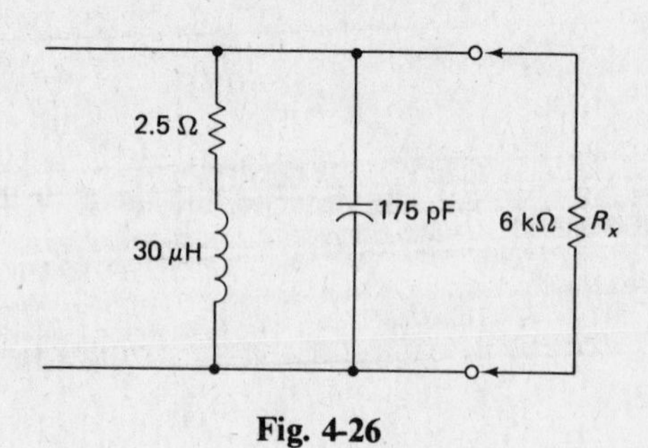

Fig. 4-26

(*a*) Applying the formula for resonant frequency,

$$\begin{aligned} f_0 &= \frac{1}{2\pi\sqrt{LC}} \\ &= \frac{1}{2\pi\sqrt{(30 \times 10^{-6})(175 \times 10^{-12})}} \\ &= 2.20 \times 10^6 \end{aligned}$$

$$\boxed{f_0 = 2.20\ \text{MHz}}$$

(*b*)

$$Q_{10} = \frac{X_L}{R}$$

First solving for X_L,

$$\begin{aligned} X_L &= 2\pi f_0 L \\ &= 2\pi(2.2 \times 10^6)(30 \times 10^{-6}) \\ &= 414.5\ \Omega \end{aligned}$$

Thus,

$$Q_{10} = \frac{X_L}{R} = \frac{414.5}{2.5}$$

$$\boxed{Q_{10} = 165.8}$$

(c)

$$Z_{10} = Q_{10} X_L$$

Substituting previously calculated values,

$$Z_{10} = (165.8)(414.5) = 68.724 \times 10^3$$

$$\boxed{Z_{10} = 68.724 \text{ k}\Omega}$$

(d)

$$Z_T = \frac{Z_{10} R_X}{Z_{10} + R_X} = \frac{(68.724 \times 10^3)(6 \times 10^3)}{(68.724 \times 10^3) + (6 \times 10^3)}$$

$$\boxed{Z_T = 5518.2 \ \Omega}$$

(e)

$$Q_T = \frac{Z_T}{X_L} = \frac{5518.2}{414.5}$$

$$\boxed{Q_T = 13.3}$$

(f)

$$\text{BW}_1 = \frac{f_0}{Q_{10}} = \frac{2.2 \times 10^6}{165.8} = 13.27 \times 10^3$$

$$\boxed{\text{BW}_1 = 13.27 \text{ kHz}}$$

(g)

$$\text{BW}_2 = \frac{f_0}{Q_T} = \frac{2.2 \times 10^6}{13.3} = 165 \times 10^3$$

$$\boxed{\text{BW}_2 = 165 \text{ kHz}}$$

4.15 Determine the necessary value for the coefficient of coupling for a double-tuned circuit in order to provide critical coupling for a situation in which the Q of the primary circuit is 60 and the Q of the secondary is 90. See Fig. 4-27.

SOLUTION

Given: $Q_P = 60$

$Q_S = 90$

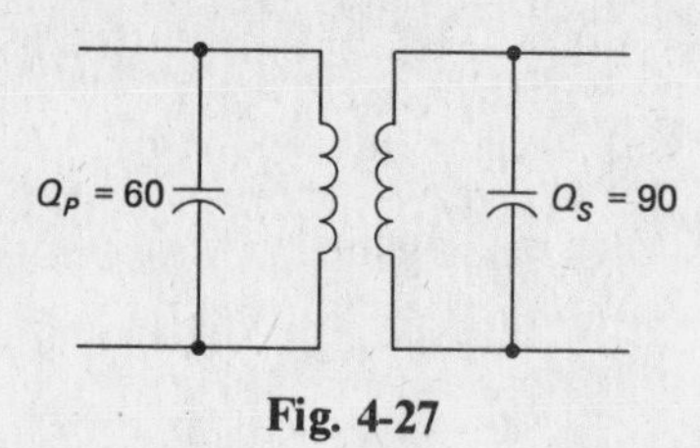

Fig. 4-27

Find: k_c

The equation for critical coupling is

$$k_c = \frac{1}{\sqrt{Q_P Q_S}} = \frac{1}{\sqrt{(60)(90)}}$$

$$\boxed{k_c = 0.0136}$$

4.16 The specifications for a double-tuned transformer to be used as a coupling network are such as to require a resonant frequency of 900 kHz and a bandwidth of 15 kHz. Both the primary and secondary of the transformer have an inductance of 200 μH. The primary circuit has a Q of 60 while the secondary has a Q of 75. See Fig. 4-28.

(*a*) Determine the capacitance required across each of the coils.
(*b*) What coefficient of coupling is required?
(*c*) Is this circuit undercoupled, critically coupled, or overcoupled?

SOLUTION

Given: $f_0 = 900$ kHz

BW = 15 kHz

$L_1 = L_2 = 200\ \mu$H

$Q_P = 60$

$Q_S = 75$

$L_1 = 200\ \mu$H, $Q_1 = 60$ $L_2 = 200\ \mu$H, $Q_2 = 75$

Fig. 4-28

Find: (*a*) C_1, C_2 (*b*) k (*c*) Type of coupling

(*a*) The required capacitance values are determined by the resonant frequency desired and the inductance values of the coils. Since both primary and secondary circuits require the same resonant frequency and since, in this case, both primary and secondary coils have the same inductance, both circuits require the same value of capacitance.

$$f_0 = \frac{1}{2\pi\sqrt{LC}}$$

$$f_0^2 = \frac{1}{4\pi^2 LC}$$

$$C = \frac{1}{4\pi^2 f_0^2 L}$$

$$C_1 = C_2 = \frac{1}{4\pi^2(900 \times 10^3)^2(200 \times 10^{-6})} = 156 \times 10^{-12}$$

$$\boxed{C_1 = C_2 = 156 \text{ pF}}$$

(*b*) The required coefficient of coupling is determined by the resonant frequency and the required bandwidth.

$$k = \frac{BW}{f_0}$$

$$= \frac{15 \times 10^3}{900 \times 10^3}$$

$$\boxed{k = 0.0167}$$

(*c*) In order to determine the type of coupling represented by the value found in (*b*), it is necessary to compare it to the value of critical coupling for this circuit.

$$k_c = \frac{1}{\sqrt{Q_P Q_S}}$$

$$= \frac{1}{\sqrt{(60)(75)}}$$

$$\boxed{k_c = 0.0149}$$

Since the value of the coefficient of coupling found in (*b*) is larger than the critical coupling value, the circuit is said to be overcoupled.

Supplementary Problems

4.17 A series circuit consists of a 0.003-μF capacitor and a 0.5-mH inductance. Calculate the resonant frequency of the combination. *Ans.* 130 kHz

4.18 Find the resonant frequency of a series combination of a 200-pF capacitor and a 400-μH inductor. *Ans.* 562 kHz

4.19 What is the resonant frequency of a series circuit consisting of a 450-pF capacitor and a 10-mH inductance? *Ans.* 75 064 Hz

4.20 What value of capacitance is required in order to provide a resonant frequency of 5 MHz when using a 40-μH inductance? *Ans.* 25.356 pF

4.21 Determine the capacitance required in order to provide a resonant frequency of 600 kHz when using a 5-mH inductance. *Ans.* 14.09 pF

4.22 A 300-pF capacitor is to be part of a series resonant circuit which is to be resonant at 1 MHz. What value of inductance should be placed in series with the capacitance? *Ans.* 84.5 μH

4.23 A series circuit resonant to 750 kHz is desired. What value of inductance should be used with a 0.004-μF capacitor? *Ans.* 11.27 μH

4.24 A 150-pF capacitor is in series with a 12-μH inductor.

(*a*) Determine the resonant frequency of the circuit.
(*b*) What is the impedance of this circuit at the resonant frequency, assuming no resistance in the circuit?
(*c*) Find the reactance of each of the components of this circuit at the resonant frequency.

Ans. (*a*) 3.753 MHz, (*b*) 0.0 Ω, (*c*) 282.83 Ω

4.25 At what frequency is a 0.05-mH inductance when it is in series with a 600-pF capacitance at resonance? The circuit also includes a 2-Ω resistance in series with the capacitor and inductor. What is the impedance of this circuit at the resonant frequency? Determine how much current would flow if a 1.5-V signal tuned to the resonant frequency of the circuit were applied to the circuit.
Ans. 919.35 kHz, 2 Ω, 0.75 A

4.26 A series circuit consists of a 0.002-μF capacitor in series with a 30-μH inductance and a 12-Ω resistance.

(*a*) What is the resonant frequency of the circuit?
(*b*) Determine the impedance of the circuit at resonance.
(*c*) Calculate the inductive and capacitive reactance at resonance.
(*d*) How much current would flow if a 500-mV source tuned to resonance were applied to the circuit?
(*e*) Calculate the voltage across each of the components for the condition described in (*d*) of this problem.

Ans. (*a*) 650 kHz; (*b*) 12 Ω; (*c*) 122.46 Ω; (*d*) 41.667 mA; (*e*) 500 mV, 5.1 V

4.27 Determine the following for the circuit shown as Fig. 4-19:

(*a*) The resonant frequency
(*b*) The total impedance at resonance
(*c*) The current that flows at resonance
(*d*) The inductive reactance at resonance
(*e*) The capacitive reactance at resonance
(*f*) The voltage across the resistor at resonance
(*g*) The voltage across the inductor at resonance
(*h*) The voltage across the capacitor at resonance

Ans. (*a*) 1.14 MHz, (*b*) 12.5 Ω, (*c*) 48 mA, (*d*) 214.776 Ω, (*e*) 214.776 Ω, (*f*) 0.6 V, (*g*) 10.309 V, (*h*) 10.309 V

600 mV at the resonant frequency of the circuit
V_0
650 pF
30 μH
12.5 Ω

Fig. 4-29

4.28 Determine the following for the circuit shown in Fig. 4-30:

(*a*) The resonant frequency
(*b*) The total impedance at resonance
(*c*) The current that flows at resonance
(*d*) The inductive reactance at resonance
(*e*) The capacitive reactance at resonance
(*f*) The voltage across the resistor at resonance
(*g*) The voltage across the inductor at resonance
(*h*) The voltage across the capacitor at resonance

Ans. (*a*) 50.355 MHz, (*b*) 2.0 Ω, (*c*) 150 mA, (*d*) 15.81 Ω, (*e*) 15.81 Ω, (*f*) 300 mV, (*g*) 2.37 V, (*h*) 2.37 V

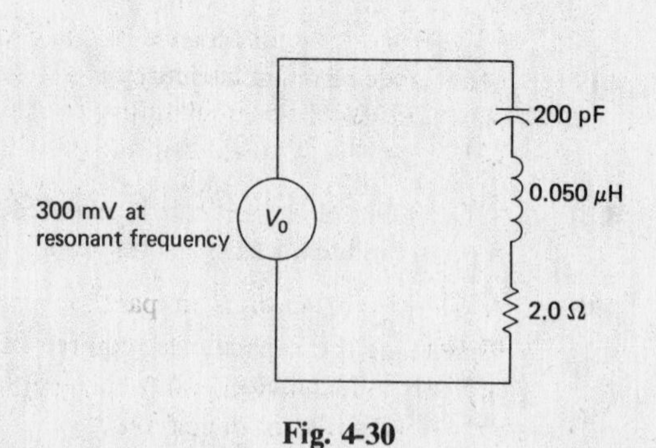

Fig. 4-30

4.29 Calculate the value of capacitance necessary in order to provide a series resonant frequency of 106 MHz when using a 20-μH inductance which has a resistance of 25 Ω. What is the Q of the circuit? Determine the bandwidth of the circuit. *Ans.* 0.1128 pF, 532, 199.248 kHz

4.30 Find the required value of capacitance for a series circuit to resonate at 700 kHz when using a 2-mH inductance and having a resistance of 18 Ω. Find the Q of the circuit and calculate the bandwidth of the circuit. *Ans.* 25.87 pF, 488.4, 1433.25 Hz

4.31 Calculate the necessary capacitance to be placed in series in order to provide a series resonance at 95 MHz. The inductance value is 25 μH. The series resistance in the circuit is 5 Ω. Find the Q of the circuit. What is the bandwidth of this circuit? *Ans.* 0.11238 pF, 2983, 31.847 kHz

4.32 A series circuit is to be resonant at 75 MHz. What value of inductance is required to be used with a 140-pF capacitance. The inductor used will have a Q of 35. What is the bandwidth of this circuit? *Ans.* 0.0322 μH, 2.14 MHz

4.33 An inductor with an inductance of 3 mH and a Q of 40 is to be made resonant to a frequency of 1 MHz by placing it in series with a capacitor. What value of capacitance should this be? Calculate the bandwidth of the circuit. How much resistance is contained in the inductor? *Ans.* 8.45 pF, 25 kHz, 471 Ω

4.34 Determine the resistance contained in a series RLC circuit which is resonant at 20 MHz and which has a Q of 30. The circuit contains an inductance of 200 μH. *Ans.* 837 Ω

4.35 Calculate the resistance of a series RLC circuit which is resonant at 850 kHz and has a Q of 40. The inductance contained in the circuit is 2 mH. What is the bandwidth of this circuit? *Ans.* 267 Ω, 21.25 kHz

4.36 Calculate the inductance contained in a series circuit which is resonant to 104 MHz and which has a bandwidth of 5.0 MHz. The circuit has a resistance of 20 Ω. *Ans.* 0.637 μH

4.37 A parallel resonant circuit consists of an inductor of 20 μH in parallel with a 400-pF capacitor. The inductor has an internal resistance of 8 Ω. A 750-mV signal at the resonant frequency of the circuit is applied across the circuit.

(*a*) Determine the resonant frequency of the circuit.
(*b*) Calculate the Q of the circuit.
(*c*) What is the impedance of this circuit at resonance?
(*d*) Determine the line current.
(*e*) Calculate the circulating current.
(*f*) Find the bandwidth of the circuit.

Ans. (*a*) 1.78 MHz, (*b*) 27.95, (*c*) 6249 Ω, (*d*) 0.12 mA, (*e*) 3.354 mA, (*f*) 63.685 kHz

4.38 A 600-pF capacitor is in parallel with a coil which has an inductance of 150 μH and a resistance of 10 Ω.

(*a*) What is the resonant frequency of this circuit?
(*b*) Find the Q of the circuit.
(*c*) How much impedance does this circuit present at resonance?
(*d*) Determine the line current when a 1.5-V source is applied. The frequency of the source is the same as the resonant frequency of the circuit.
(*e*) Calculate the circulating current when the 1.5-V signal is applied.
(*f*) Find the bandwidth of the circuit.

Ans. (*a*) 530 kHz, (*b*) 49.9, (*c*) 24.9 kΩ, (*d*) 0.06 mA, (*e*) 2.99 mA, (*f*) 10.62 kHz

4.39 A 350-pF capacitor is in parallel with a coil which has an inductance of 200 μH and a resistance of 5 Ω.

(*a*) Determine the resonant frequency of this circuit.
(*b*) Find the Q of the circuit.
(*c*) How much impedance does this circuit present at resonance?
(*d*) Determine the line current when a 1.5-V source is applied. The frequency of the source is the same as the resonant frequency of the circuit.
(*e*) Calculate the circulating current when the 1.5-V signal is applied.
(*f*) Find the bandwidth of the circuit.

Ans. (*a*) 601.85 kHz, (*b*) 151.2, (*c*) 114.297 kΩ, (*d*) 13.12 μA, (*e*) 1.98 mA, (*f*) 3.98 kHz

4.40 A 0.01-μF capacitor is in parallel with a coil which has an inductance of 0.025 mH and a resistance of 10 Ω.

(*a*) What is the resonant frequency of this circuit?
(*b*) Find the Q of the circuit.
(*c*) How much impedance does this circuit present at resonance?
(*d*) Determine the line current when a 3.5-V source is applied. The frequency of the source is the same as the resonant frequency of the circuit.
(*e*) Calculate the circulating current when the 3.5-V signal is applied.
(*f*) Find the bandwidth of the circuit.

Ans. (*a*) 318.47 kHz, (*b*) 5.0, (*c*) 250 Ω, (*d*) 14 mA, (*e*) 70 mA, (*f*) 6.37 kHz

4.41 An external resistance load is installed in parallel with an RLC circuit as shown in Fig. 4-31.

(*a*) Find the frequency at which the circuit is resonant.
(*b*) What was the Q of the circuit prior to installation of the external load?
(*c*) Find the impedance of the circuit at resonance prior to installing the external resistance.
(*d*) Determine the impedance of the total circuit at resonance after the external resistance is installed.
(*e*) Find the Q of the circuit after installing the external resistance.
(*f*) Calculate the bandwidth of the circuit prior to installation of the external resistance.
(*g*) What is the bandwidth of the circuit after installing the external load?

Ans. (*a*) 4.914 MHz, (*b*) 43.2, (*c*) 4665.6 Ω, (*d*) 3558.7 Ω, (*e*) 32.95, (*f*) 113.75 kHz, (*g*) 149.135 kHz

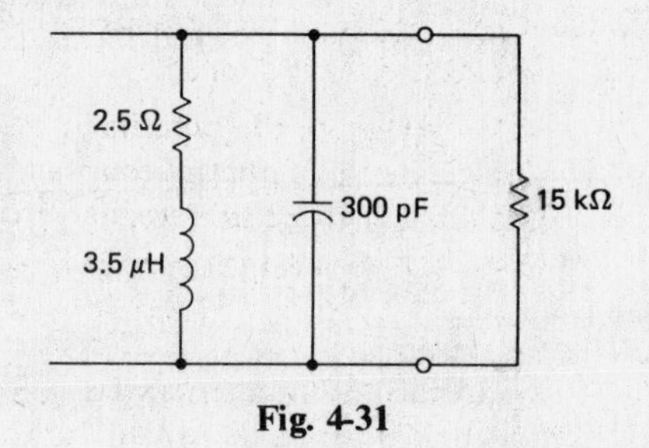

Fig. 4-31

4.42 An external resistance load is installed in parallel with an RLC circuit as shown in Fig. 4-32.

(*a*) Find the frequency at which the circuit is resonant.
(*b*) What was the Q of the circuit prior to installation of the external load.
(*c*) Find the impedance of the circuit at resonance prior to installing the external resistance.
(*d*) Determine the impedance of the total circuit at resonance after the external resistance is installed.
(*e*) Find the Q of the circuit after installing the external resistance.
(*f*) Calculate the bandwidth of the circuit prior to installation of the external resistance.
(*g*) What is the bandwidth of the circuit after installing the external load?

Ans. (*a*) 750.64 kHz, (*b*) 94.28, (*c*) 44.443 kΩ, (*d*) 8163.2 Ω, (*e*) 17.32, (*f*) 7962 Hz, (*g*) 43.34 kHz

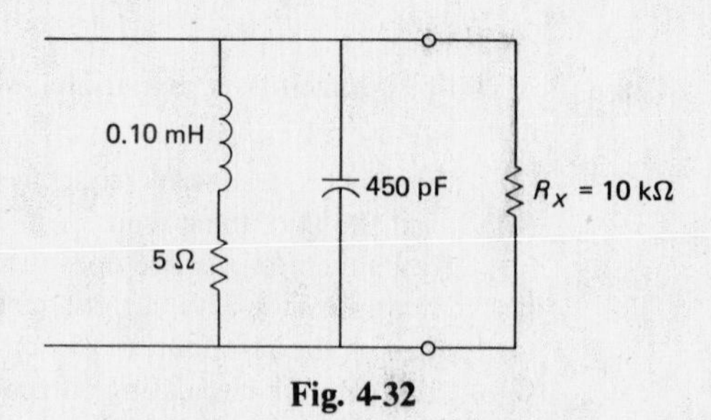

Fig. 4-32

4.43 An external resistance load is installed in parallel with an RLC circuit as shown in Fig. 4-33.

(*a*) Find the frequency at which the circuit is resonant.
(*b*) What was the Q of the circuit prior to installation of the external load?
(*c*) Find the impedance of the circuit at resonance prior to installing the external resistance.
(*d*) Determine the impedance of the total circuit at resonance after the external resistance is installed.
(*e*) Find the Q of the circuit after installing the external resistance.
(*f*) Calculate the bandwidth of the circuit prior to installation of the external resistance.
(*g*) What is the bandwidth of the circuit after installing the external load?

Ans. (*a*) 7.12 MHz, (*b*) 27.95, (*c*) 625 Ω, (*d*) 340.9 Ω, (*e*) 15.25, (*f*) 254.74 kHz, (*g*) 466.885 kHz

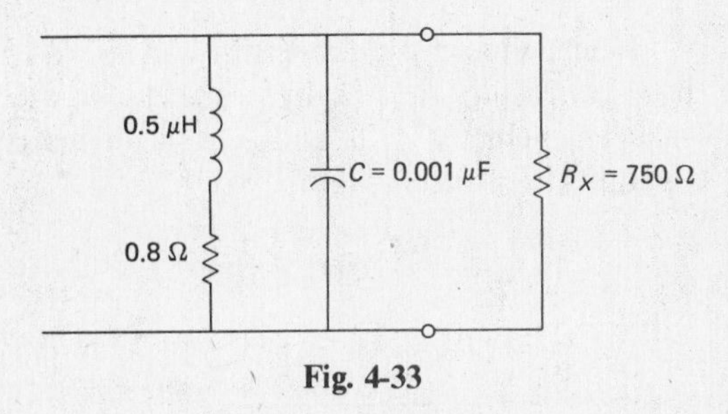

Fig. 4-33

4.44 Determine the necessary coefficient of coupling in order to provide critical coupling for a double-tuned situation in which the Q of the primary circuit is 25 and the Q of the secondary circuit is 40.
Ans. 0.0316

4.45 The specifications for a double-tuned transformer to be used as a coupling network are such as to require a resonant frequency of 700 kHz and a bandwidth of 25 kHz. Both the primary and secondary of the transformer have an inductance of 75 μH. The primary circuit has a Q of 20 while the secondary has a Q of 15.

(*a*) Determine the capacitance required across each of the coils.
(*b*) What coefficient of coupling is required in order to meet these specifications?
(*c*) Is this circuit undercoupled, critically coupled, or overcoupled?

Ans. (*a*) 689.96 μF, (*b*) .0357, (*c*) undercoupled

Chapter 5

RF Amplifiers and Oscillators

When choosing a transistor for use as the active device in RF amplifiers or oscillators, only those transistors specifically designed for use at RF should be considered, since the various forward gain parameters α, β, μ, and g_m roll off toward zero as frequency is increased beyond the frequency capability of a specific unit. See Figs. 5-1 and 5-2.

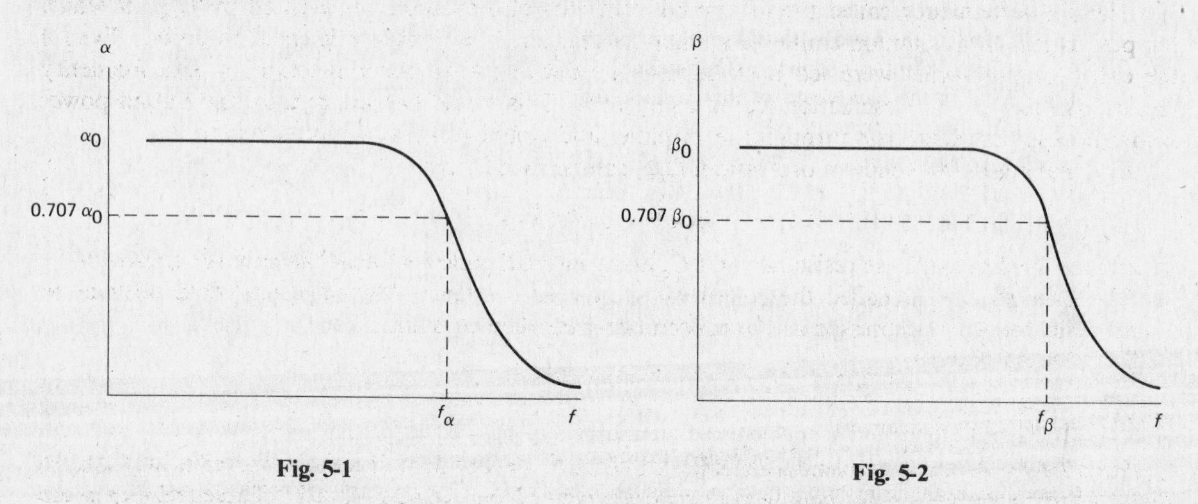

Fig. 5-1 **Fig. 5-2**

Manufacturers' data for transistors intended for use at RF frequencies include the following parameters:

$$f_\alpha, \quad f_\beta, \quad f_T$$

as well as α_0 and β_0, the low-frequency α and β for the transistor.

f_α, the frequency at which the α of the transistor has dropped to 0.707 of its low-frequency value, α_0, is called the *alpha cutoff frequency* of the transistor. The alpha cutoff frequency is also called the 3-dB frequency of the transistor because 3-dB represents half power, and the power gain of the transistor is directly proportional to the square of α. Thus if α is down by 0.707, power gain is down by $(0.707)^2$, which is 1/2.

The frequency at which β is down by 0.707 of its low-frequency value is called the *beta cutoff frequency*, f_β.

f_β is a lower frequency than f_α for a particular transistor, the relationship being approximated by*

$$f_\beta = 0.8(1 - \alpha)f_\alpha$$

The more exact relationship is

$$f_\beta = K_\theta(1 - \alpha)f_\alpha$$

where K_θ is a transistor parameter which ranges from 0.5 to 1.0, with most transistors having a K_θ between 0.8 and 1.0.

f_T is the frequency at which β has fallen to 1.0 and is frequently referred to as the *gain bandwidth product* of the transistor because it can be found by multiplying β by the frequency at which it is being measured.

---

* Motorola Semiconductor Products Tech. Info. Note AN-139 by Roy Hejhall.

f_T is also the product of β_0, the low-frequency value of β, and f_β, the β cutoff frequency:*

$$f_T = \beta_0 \times f_\beta$$

By a series of approximations, it can be shown that the gain bandwidth product can be approximated by f_α.

$$f_T \approx f_\alpha$$

THE MAXIMUM FREQUENCY OF OSCILLATION

There is a frequency called the maximum frequency of oscillation, represented by f_{max}, at which the power gain of a common-emitter amplifier circuit using the transistor is equal to unity. Even if the entire output signal were fed back to the input with the amplifier functioning at a frequency higher than the f_{max}, the oscillation would damp out. This would happen because the output power would be less with each pass through the amplifier and would eventually approach zero.

The maximum frequency of oscillation, f_{max}, can be found from

$$f_{max} = \sqrt{f_T / 8\pi r'_b C_c}$$

where r'_b is the internal base resistance and C_c is the internal collector capacitance of the transistor.

The product $r'_b C_c$ is called the *collector-to-base time constant*. Manufacturers' data relevant to r'_b and C_c is frequently expressed as the collector-to-base time constant.

NEUTRALIZATION

Because of the great tendency for RF amplifiers to be unstable (break into oscillation) due to signal feedback, techniques for counteracting this tendency have been developed. These techniques are referred to as neutralization techniques.

Neutralization involves providing a second feedback path to cancel the other feedback. The neutralization feedback should be 180° out of phase with the other feedback for cancellation to occur.

A simple example of this technique is shown in Fig. 5-3, the neutralization capacitor C_N providing a means of feeding back the signal which was shifted by 180° by the transformer.

Figure 5-4 depicts another means of obtaining a 180° phase shifted signal to feed back to the input. This technique shown in Fig. 5-4 is called *collector neutralization.*

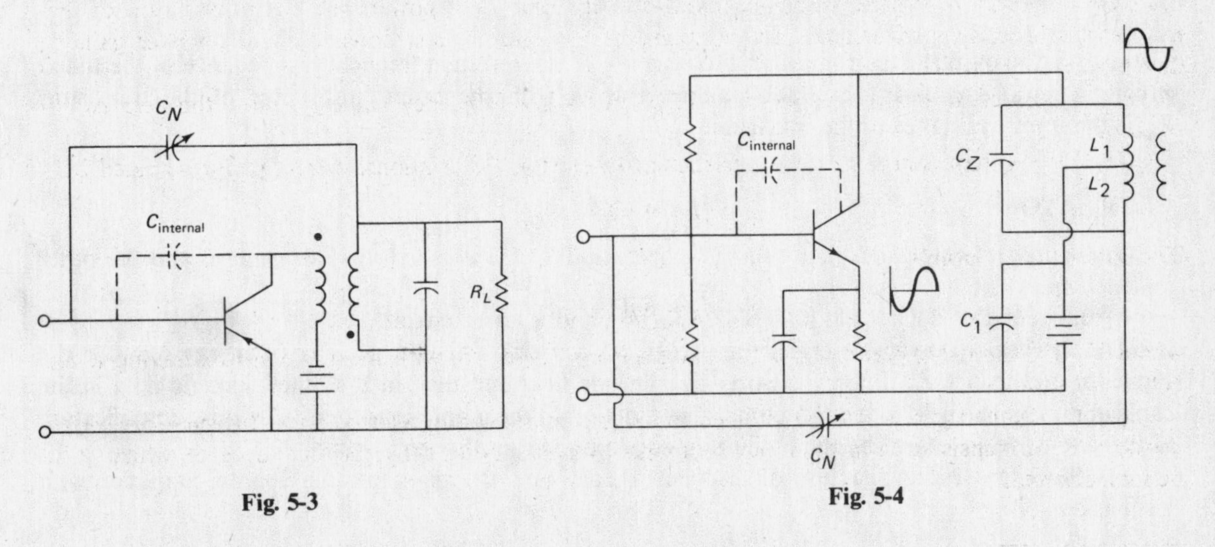

Fig. 5-3 Fig. 5-4

* Motorola Semiconductor Products Tech. Info. Note AN-139 by Roy Hejhall.

Figure 5-5 can be shown to be an equivalent circuit representative of the neutralization portion of the circuit. For the most effective neutralization, the bridge of Fig. 5-5 should be balanced. The balance equation is

$$\frac{C_N}{C_{\text{internal}}} = \frac{L_1}{L_2}$$

where C_{internal} includes external stray capacitance from collector to base, as well as the internal capacitance of the active device.

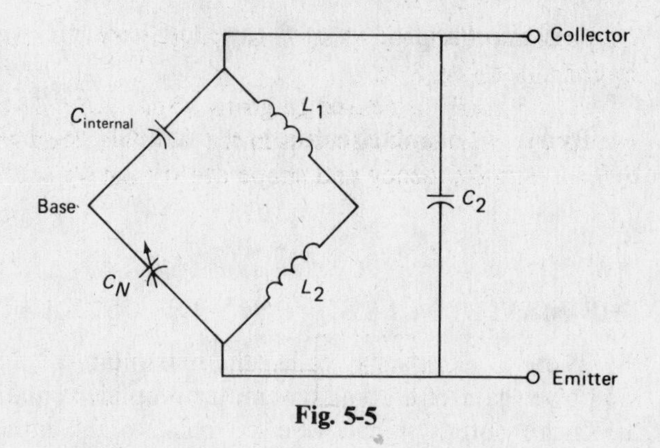

Fig. 5-5

RF COUPLING

Coupling between RF amplifiers is frequently accomplished by *tuned impedance coupling* as shown in Fig. 5-6. This technique is usually preferred over the *resistance-capacitance coupling* technique which is quite popular at audio frequencies.

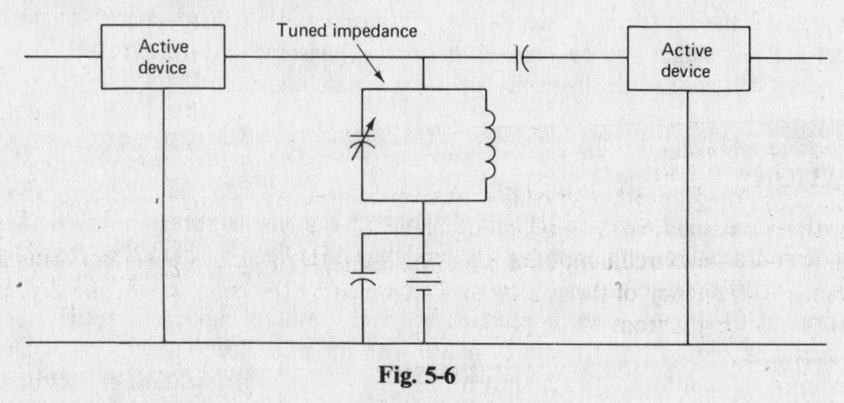

Fig. 5-6

The voltage gain of a transistor amplifier which is tuned impedance coupled is

$$A_V = -\beta \frac{Z_0}{R_{\text{in}}}$$

where Z_0 is the output circuit impedance consisting of the resonant frequency impedance of the tuned circuit in parallel with any external load as well as with the output resistance of the transistor. R_{in} is the input resistance of the transistor.

OSCILLATORS

There are two ways to explain oscillator operation. One is called *flywheel analysis* and the other *feedback analysis.*

Most oscillators consist of an amplifier portion and a tuned circuit.

The flywheel approach describes the oscillation as being due to a transfer of energy between the capacitor and inductor. Energy is stored first in one field and then in the other: electric field in the capacitor, magnetic field in the inductor. The function of the amplifier is to provide only amplification of the sine wave generated by the oscillation occurring within the tank circuit.

Feedback analysis makes use of feedback theory and the gain formula for an amplifier with feedback.

$$A' = \frac{A}{1 - AB}$$

where A' is the total gain, A is the gain of the amplifier without feedback, and B is the feedback fraction (see Fig. 5-7).

If $1 - AB$ is caused to go to zero, $AB = 1$ and the system provides its own input signal and oscillates. The tank circuit is in the feedback loop and determines at which frequency $AB = 1$, thus determining frequency and shape of the signal generated.

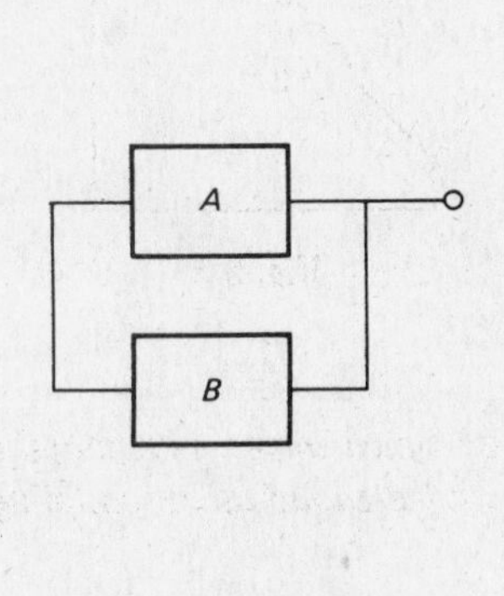

Fig. 5-7

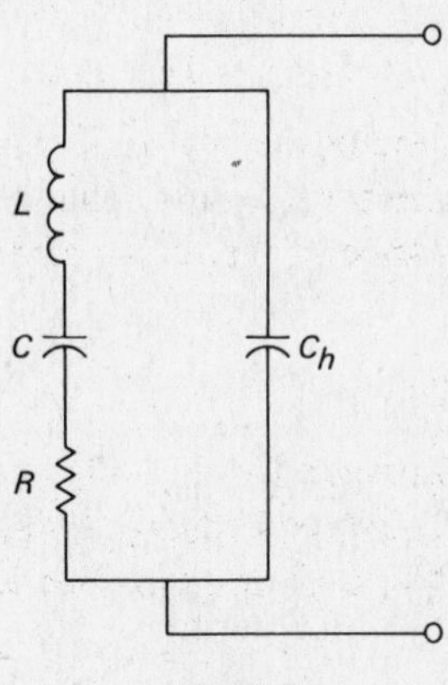

Fig. 5-8

CRYSTAL OSCILLATORS

Because quartz and some other crystal materials exhibit a property known as the *piezoelectric effect*, they can be represented by the equivalent circuit of Fig. 5-8.

Note that this equivalent circuit contains L's and C's and therefore has a natural resonant frequency. The natural resonant frequency of the crystal is determined by the geometry of the crystal, which can be controlled when cutting it from its large block.

Solved Problems

5.1 The alpha cutoff frequency for a particular transistor is 900 MHz. Determine the beta cutoff frequency. The low-frequency common-base forward current gain is equal to 0.98.

SOLUTION

Given: $f_\alpha = 900$ MHz
$\alpha_0 = 0.98$

Find: f_β

The relationship between f_α and f_β for an RF transistor can be approximated by

$$\begin{aligned} f_\beta &= 0.8(1 - \alpha)f_\alpha \\ &= 0.8(1 - 0.98)(900 \times 10^6) \\ &= 0.8(0.02)(900 \times 10^6) \\ &= 14.4 \times 10^6 \end{aligned}$$

$$\boxed{f_\beta = 14.4 \text{ MHz}}$$

5.2 A transistor has an alpha cutoff frequency of 850 MHz and a low-frequency alpha (α_0) of 0.985. Determine the beta cutoff frequency for the device.

SOLUTION

Given: $f_\alpha = 850$ MHz
$\alpha_0 = 0.985$

Find: f_β

$$f_\beta = K_\theta(1 - \alpha)f_\alpha$$

Assume $K_\theta = 0.8$. Thus,

$$\begin{aligned} f_\beta &= 0.8(1 - 0.985)(850 \times 10^6) \\ &= 0.8(0.015)(850 \times 10^6) \\ &= 10.2 \times 10^6 \end{aligned}$$

$$\boxed{f_\beta = 10.2 \text{ MHz}}$$

5.3 An RF transistor has a beta cutoff frequency (f_β) of 700 kHz and a low-frequency beta (β_0) of 60. Determine the gain bandwidth product for this device.

SOLUTION

Given: $f_\beta = 700$ kHz
$\beta_0 = 60$

Find: f_T

The gain bandwidth product of a transistor can be found by multiplying the low-frequency beta of the transistor by the beta cutoff frequency:

$$\begin{aligned} f_T &= \beta_0 \times f_\beta \\ &= 60 \times (700 \times 10^3) \\ &= 42 \times 10^6 \end{aligned}$$

$$\boxed{f_T = 42 \text{ MHz}}$$

5.4 Calculate (*a*) the alpha cutoff frequency, (*b*) the beta cutoff frequency, and (*c*) the low-frequency value of alpha for a transistor having a gain bandwidth product f_T of 1200 MHz and a low-frequency beta of 90.

SOLUTION

Given: $f_T = 1200$ MHz
$\beta_0 = 90$

Find: (*a*) f_α (*b*) f_β (*c*) α_0

(*a*) By a series of approximations involving allowing α_0 and K_θ to be approximated by 1.0, f_T can be shown to be approximately equal to f_α.

$$f_\alpha \approx f_T$$

$$\boxed{f_\alpha \approx 1200 \text{ MHz}}$$

(*b*) The relationship between f_β, β_0, and f_T is

$$f_T = \beta_0 \times f_\beta$$

Solving for f_β,

$$f_\beta = \frac{f_T}{\beta_0} = \frac{1200 \times 10^6}{90} = 13.33 \times 10^6$$

$$\boxed{f_\beta = 13.33\ \text{MHz}}$$

(*c*) The standard relationship between α and β is

$$\alpha_0 = \frac{\beta_0}{1+\beta_0} = \frac{90}{1+90} = \frac{90}{91}$$

$$\boxed{\alpha_0 = 0.989}$$

5.5 Calculate the maximum frequency of oscillation of a transistor which has a gain bandwidth product of 65 MHz and a collector-to-base time constant $r'_b C_c$ of 15 picoseconds (ps).

SOLUTION

Given: $f_T = 65$ MHz
$r'_b C_c = 15$ ps

Find: f_{max}

Using the equation relating f_{max}, f_T, and $r'_b C_c$,

$$f_{max} = \sqrt{\frac{f_T}{8\pi r'_b C_c}} = \sqrt{\frac{65 \times 10^6}{8\pi(15 \times 10^{-12})}} = 415.34 \times 10^6$$

$$\boxed{f_{max} = 415.34\ \text{MHz}}$$

5.6 The maximum frequency of oscillation for a transistor is 1400 MHz and the collector-to-base time constant $r'_b C_c$ is 50 ps. Calculate the gain bandwidth product for this transistor.

SOLUTION

Given: $f_{max} = 1400$ MHz
$r'_b C_c = 50$ ps

Find: f_T

The equation relating f_T, f_{max}, and $r'_b C_c$ is

$$f_{max} = \sqrt{\frac{f_T}{8\pi r'_b C_c}}$$

Solving for f_T, first square both sides of the equation:

$$f^2_{max} = \frac{f_T}{8\pi r'_b C_c}$$

Now multiply both sides of the equation by $8\pi r'_b C_c$:

$$f^2_{max}\, 8\pi r'_b C_c = \frac{f_T}{8\pi r'_b C_c} \times 8\pi r'_b C_c$$

$$f^2_{max}\, 8\pi r'_b C_c = f_T$$

Rearranging some of the terms for convenience,

$$f_T = 8\pi r'_b C_c f^2_{max}$$

Substituting numerical values,

$$f_T = 8\pi(50 \times 10^{-12})(1400 \times 10^6)^2$$
$$= 2461.76 \times 10^6$$

$$\boxed{f_T = 2461.76 \text{ MHz}}$$

5.7 What is the collector-to-base time constant $r'_b C_c$ for a transistor that has a maximum frequency of oscillation of 650 MHz and a gain bandwidth product of 45 MHz?

SOLUTION

Given: $f_{max} = 650$ MHz
$f_T = 45$ MHz

Find: $r'_b C_c$

The relationship between f_{max}, f_T, and $r'_b C_c$ is

$$f_{max} = \sqrt{\frac{f_T}{8\pi r'_b C_c}}$$

Solving for $r'_b C_c$, first square both sides of the equation:

$$f^2_{max} = \frac{f_T}{8\pi r'_b C_c}$$

Multiply both sides of the equation by $r'_b C_c$, so that

$$r'_b C_c f^2_{max} = \frac{f_T}{8\pi r'_b C_c} r'_b C_c$$
$$= \frac{f_T}{8\pi}$$

Now divide both sides of the equation by f^2_{max};

$$\frac{r'_b C_c f^2_{max}}{f^2_{max}} = \frac{f_T}{8\pi f^2_{max}}$$

$$r'_b C_c = \frac{f_T}{8\pi f^2_{max}}$$

Substituting numerical values,

$$r'_b C_c = \frac{45 \times 10^6}{8\pi(650 \times 10^6)^2}$$
$$= 4.24 \times 10^{-12}$$

$$\boxed{r'_b C_c = 4.24 \text{ ps}}$$

5.8 The circuit shown as Fig. 5-9 has an internal and stray wiring capacitance represented by C_{internal} of 50 pF. If $L_1 = 60$ mH and $L_2 = 35$ mH, determine the value to which the neutralizing capacitor C_N should be set so as to neutralize C_{internal}.

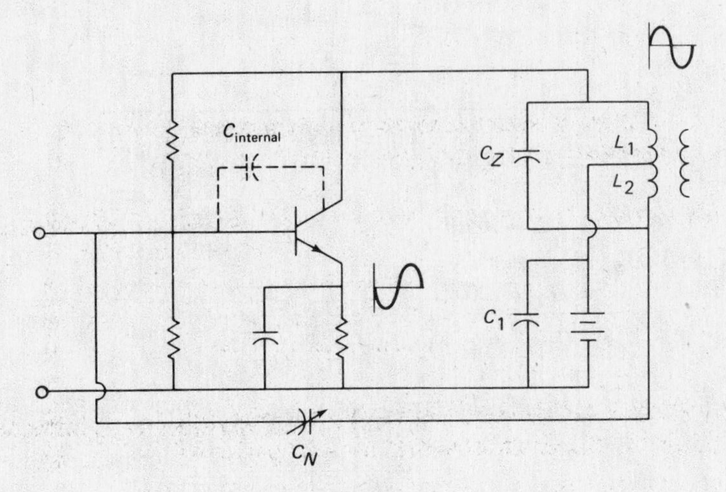

Fig. 5-9

SOLUTION

Given: $C_{\text{internal}} = 50$ pF
$L_1 = 60$ mH
$L_2 = 35$ mH

Find: C_N

That portion of the circuit involved with neutralization of the circuit of Fig. 5-9 can be represented by a bridge circuit whose balance equation is

$$\frac{C_N}{C_{\text{internal}}} = \frac{L_1}{L_2}$$

Thus,

$$\frac{C_N}{50 \times 10^{-12}} = \frac{60 \times 10^{-3}}{35 \times 10^{-3}}$$

$$C_N = \frac{60 \times 10^{-3}}{35 \times 10^{-3}} (50 \times 10^{-12})$$
$$= 8.57 \times 10^{-11} \quad \text{or} \quad 85.7 \times 10^{-12}$$

$$\boxed{C_N = 85.7 \text{ pF}}$$

5.9 A common-emitter amplifier as shown in Fig. 5-10 makes use of a transistor having a β of 60. The input impedance R_{in} of the transistor is 350 Ω. The capacitor has a capacitance of 250 pF while the coil has an inductance of 50 μH. The resistance of the coil is 25 Ω and the external load $R_L = 2000\ \Omega$. Determine (*a*) the resonant frequency of the circuit, (*b*) the Q of the coil, (*c*) the total impedance of the output circuit, and (*d*) the voltage gain of the amplifier at the resonant frequency.

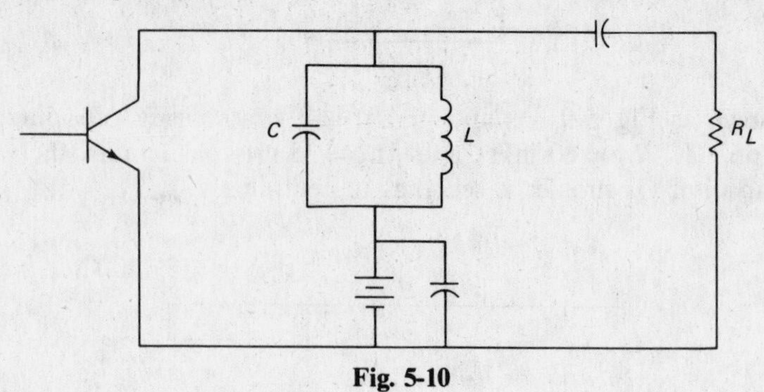

Fig. 5-10

SOLUTION

Given: $\beta = 60$ $L = 50\ \mu\text{H}$

$R_{in} = 350\ \Omega$ $R = 25\ \Omega$

$C = 250\ \text{pF}$ $R_L = 2000\ \Omega$

Find: (*a*) f_0 (*b*) Q_1 (*c*) Z_T (*d*) A_V

(*a*)

$$f_0 = \frac{1}{2\pi\sqrt{LC}}$$

$$= \frac{1}{2\pi\sqrt{(50 \times 10^{-6})(250 \times 10^{-12})}}$$

$$= 1.42 \times 10^6$$

$$\boxed{f_0 = 1.42\ \text{MHz}}$$

(*b*)

$$Q_1 = \frac{X_L}{R}$$

$$= \frac{2\pi f_0 L}{R}$$

$$= \frac{2\pi(1.42 \times 10^6)(50 \times 10^{-6})}{25}$$

$$\boxed{Q_1 = 17.84}$$

(*c*) The total impedance at the resonant frequency of the output circuit is equal to the parallel combination of the impedance of the tuned circuit, the load resistor R_L, and the output resistance of the transistor:

$$Z_T = Z_{10} \| R_L \| R_{out}$$

Solving for Z_{10} first,

$$Z_{10} = Q_1 X_L$$

$$= Q_1(2\pi f_0 L)$$

$$= (17.84)(2\pi)(1.42 \times 10^6)(50 \times 10^{-6})$$

$$= 7.95 \times 10^3$$

$$= 7.95\ \text{k}\Omega$$

Assuming the output resistance of the transistor to be large enough to be neglected,

$$Z_T = Z_{10} \| R_L$$
$$= \frac{Z_{10} R_L}{Z_{10} + R_L}$$
$$= \frac{(7.95 \times 10^3)(2 \times 10^3)}{(7.95 \times 10^3) + (2 \times 10^3)}$$
$$= \frac{15.9 \times 10^6}{9.95 \times 10^3}$$

$$\boxed{Z_T = 1.6\ \text{k}\Omega}$$

(*d*)
$$A_V = -\beta \frac{Z_T}{R_{in}}$$
$$= -60 \frac{1600}{350}$$

$$\boxed{A_V = -274.3}$$

5.10 In Fig. 5-11 the gain of the amplifier without feedback A is 12. Determine the required value of the feedback fraction B in order for this circuit to oscillate.

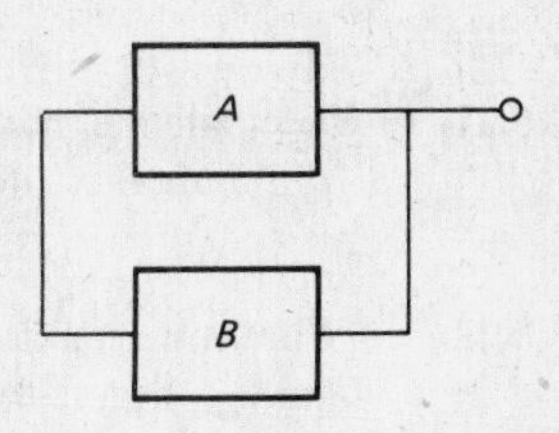

Fig. 5-11

SOLUTION

Given: $A = 12$

Find: B

In order for oscillation to occur, the product AB must equal 1.0. Thus,

$$AB = 1$$
$$12(B) = 1$$
$$B = \tfrac{1}{12}$$

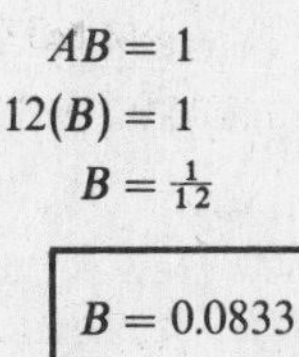

$$\boxed{B = 0.0833}$$

Supplementary Problems

5.11 Why is it necessary when designing RF amplifiers to use only transistors designed for RF?
Ans. α, β, μ, and g_m roll off toward zero at RF for transistors not specifically designed for use at these frequencies.

5.12 What is meant by f_α, f_β, and f_T? What is their significance? *Ans.* See beginning of Chapter 5.

5.13 The alpha cutoff frequency of a transistor is 1050 MHz. The low-frequency alpha of this transistor is 0.987. Calculate the beta cutoff frequency of this device. *Ans.* 10.92 MHz

5.14 Determine the beta cutoff frequency for a transistor having an alpha cutoff frequency of 750 MHz and a low-frequency alpha of 0.975. *Ans.* 15 MHz

5.15 A transistor having a low-frequency alpha of 0.983 has a beta cutoff frequency of 8 MHz. What is the alpha cutoff frequency for this device? *Ans.* 588.235 MHz

5.16 Find the alpha cutoff frequency for a transistor that has a beta cutoff frequency of 5 MHz and which has a low-frequency alpha of 0.978. *Ans.* 284.09 MHz

5.17 An RF transistor has a beta cutoff frequency of 600 kHz and a low-frequency beta of 40. Determine the gain bandwidth product for this device. *Ans.* 24 MHz

5.18 Determine the alpha cutoff frequency, the beta cutoff frequency, and the low-frequency alpha for a transistor having a gain bandwidth product of 2000 MHz and a low-frequency beta of 100.
Ans. 2000 MHz, 20 MHz, 0.990

5.19 What is the significance of f_{max}, the maximum frequency of oscillation?
Ans. It is the frequency at which the power gain equals unity.

5.20 A transistor has a maximum frequency of oscillation of 900 MHz and a gain bandwidth product of 250 MHz. Determine the collector-to-base time constant of this transistor. *Ans.* 12.287 ps

5.21 Calculate the maximum frequency of oscillation of a transistor that has a collector-to-base time constant ($r'_b C_c$) of 10 ps, if the gain bandwidth product of the device is 600 MHz. *Ans.* 1545.5 MHz

5.22 What is the maximum frequency of oscillation of a transistor having a gain bandwidth product of 50 MHz and a collector-to-base time constant ($r'_b C_c$) of 12 ps? *Ans.* 407.3 MHz

5.23 A transistor has a maximum frequency of oscillation of 800 MHz and a collector-to-base time constant ($r'_b C_c$) of 25 ps. Calculate the gain bandwidth product of this device. *Ans.* 401.92 MHz

5.24 What is the function of neutralization? How is this function realized?
Ans. Neutralization reduces instability. It provides a second feedback path out of phase by 180°.

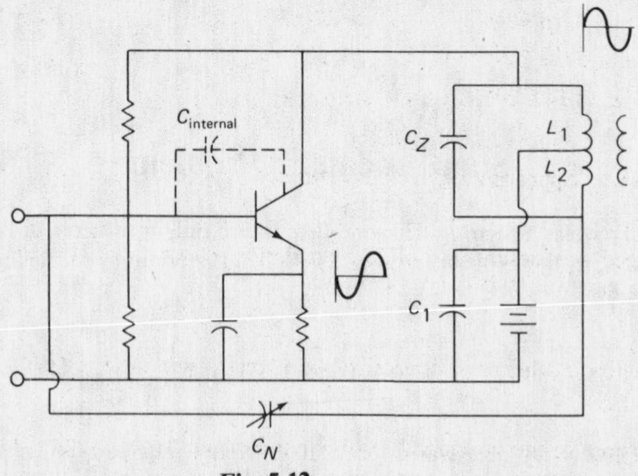

Fig. 5-12

5.25 The circuit shown as Fig. 5-12 has an internal and stray wiring capacitance of 25 pF. If $L_1 = 75$ mH and $L_2 = 125$ mH, determine to what value the neutralizing capacitor C_N should be set so as to neutralize C_{internal}. *Ans.* 15 pF

5.26 The common-emitter amplifier shown as Fig. 5-13 uses a transistor which has a β of 85. The input impedance of the transistor is 600 Ω. The capacitor has a capacitance of 400 pF and the coil has an inductance of 75 μH. The resistance of the coil is 40 Ω. The external load R_L has a resistance of 3000 Ω.

(*a*) Calculate the resonant frequency of the tuned circuit.
(*b*) What is the Q of the coil?
(*c*) Determine the impedance of the output circuit at resonance.
(*d*) Find the gain of the amplifier at resonance.

Ans. (*a*) 919 kHz, (*b*) 10.82, (*c*) 1828.6 Ω, (*d*) −259

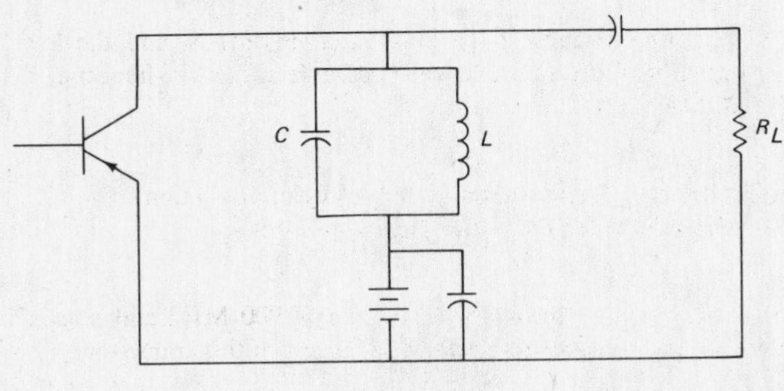

Fig. 5-13

5.27 Again consider the common-emitter amplifier shown as Fig. 5-13. Determine the resonant frequency, the Q of the coil, the impedance of the output circuit at resonance, and the gain of the amplifier at resonance for a situation in which the following component values are in use.

Transistor	*Capacitor*	*Coil*	*Load*
$\beta = 75$	$C = 175$ pF	$L = 65$ μH	$R_L = 4000$ Ω
$R_{\text{in}} = 575$ Ω		$R = 60$ Ω	

Ans. 1.493 MHz, 10.157, 2430 Ω, −317

5.28 Repeat Problem 5.27 for the following component values:

Transistor	*Capacitor*	*Coil*	*Load*
$\beta = 95$	$C = 200$ pF	$L = 70$ μH	$R_L = 3500$ Ω
$R_{\text{in}} = 800$ Ω		$R = 45$ Ω	

Ans. 1.346 MHz, 13.147, 2413.78 Ω, −286.6

5.29 Discuss the flywheel approach and how it describes how oscillators work. *Ans.* See page 86.

5.30 How does the feedback description of how oscillation works differ from the flywheel approach? *Ans.* See pages 86–87.

5.31 What is the purpose of the amplifier according to the flywheel approach? *Ans.* Only to amplify.

5.32 What is the purpose of the tank circuit in the feedback approach to oscillation theory? *Ans.* To determine frequency and waveshape.

5.33 Determine the value of B in order for oscillation to occur in the circuit as shown as Fig. 5-14, if $A = 40$. *Ans.* 0.025

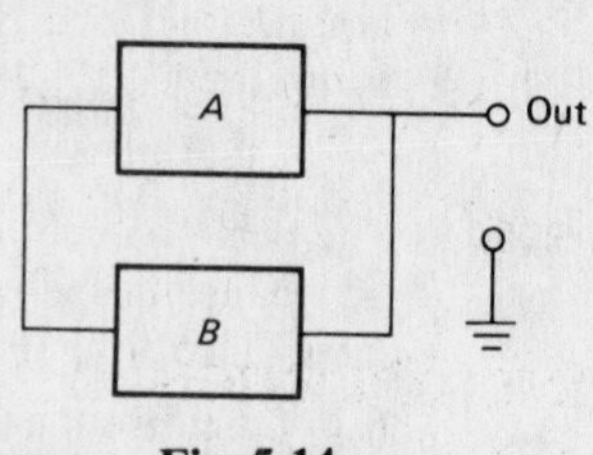

Fig. 5-14

5.34 What property makes crystals usable as frequency-determining devices in an oscillator? *Ans.* Piezoelectric effect.

5.35 What determines the natural resonant frequency of a crystal? *Ans.* The geometry of the crystal.

Chapter 6

Transmission Lines

INTRODUCTION

Electromagnetic waves travel in free space at 3×10^8 meters/s or 186 000 miles/s. In other than free space, electromagnetic waves travel a bit slower; however, as a first approximation the speed in free space can be assumed to be the speed on a transmission line. If a signal varies so rapidly or the line is so long that before the leading edge of the signal reaches the end of the transmission line the incoming signal undergoes an appreciable change, we must consider how the transmission line affects the signal.

A PULSE ON A TRANSMISSION LINE

Consider a pulse launched on a 100-meter transmission line. The pulse is to be of such short duration that the leading edge does not arrive at the load end of the line before the trailing edge of the pulse has left the generator. The amount of time for the signal to travel the length of the transmission line can be calculated as follows:

$$\frac{100 \text{ meters}}{3 \times 10^8 \text{ meters/s}} = 33.3 \times 10^{-8} \text{ s} \qquad \text{or} \qquad 333 \text{ ns}$$

If the pulse width is less than 333 ns in duration, the voltage at the generator will have returned to zero before the leading edge of the pulse arrives at the load. Figure 6-1 describes this situation.

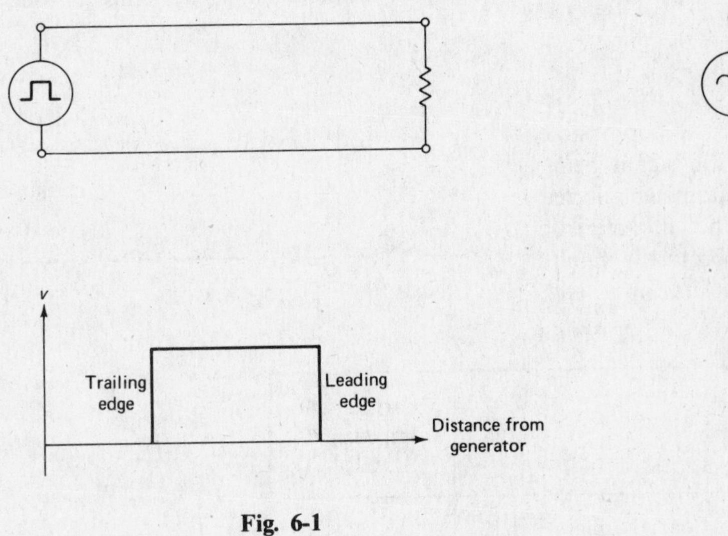

Fig. 6-1

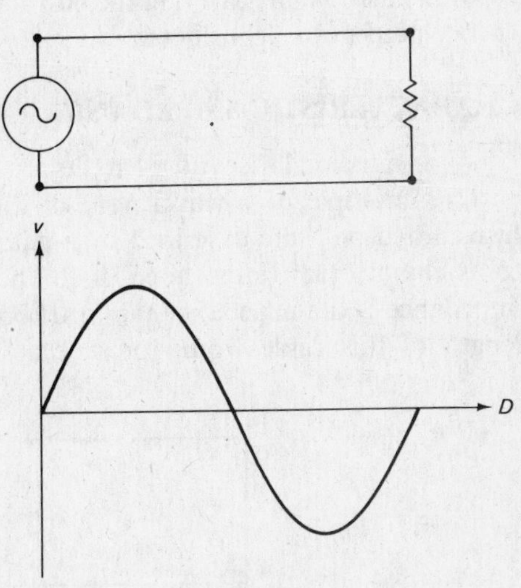

Fig. 6-2

A SINE WAVE ON A TRANSMISSION LINE

Now consider a *sine wave* traveling on a transmission line whose length is appreciable compared to the wavelength of the sine wave. The wavelength of the sine wave is defined as the distance traveled by an electromagnetic signal during one cycle of the sine wave. Figure 6-2 shows a sine wave on a transmission line. The sine wave in this case has a wavelength equal to the physical length of the line.

Since the instantaneous value of a sine wave is constantly changing, we can imagine the ramifications involved when the length of the transmission line is of the same order of magnitude as the wavelength of the signal traveling on it. The instantaneous voltage anywhere on the line will be different from that elsewhere on the line.

The reason this situation has not been encountered with transmission lines which were much shorter than a wavelength is that although the voltage everywhere on the line was different, the difference was very minute since the speed of propagation was much greater than the rate of change of the signal. Consider the wavelength of a 60-Hz sine wave:

$$f\lambda = 3 \times 10^8 \text{ meters/s}$$
$$60\lambda = 3 \times 10^8 \text{ meters/s}$$
$$\lambda = \frac{3 \times 10^8}{60} = 0.05 \times 10^8$$
$$\boxed{\lambda = 5 \times 10^6 \text{ meters}}$$

or

$$f\lambda = 186\,000 \text{ miles/s}$$
$$60\lambda = 186\,000$$
$$\lambda = \frac{186\,000}{60}$$
$$\boxed{\lambda = 3100 \text{ miles}}$$

Thus we see that the wavelength of a 60-Hz signal is 5 million meters or 3100 miles. Unless the transmission line carrying the 60-Hz signal is an appreciable part of this distance, transmission-line theory need not be considered.

CHARACTERISTIC IMPEDANCE

Transmission lines whose lengths are an appreciable part or multiple of a wavelength of the signal being transmitted on it are described by a parameter referred to as characteristic impedance Z_0. The characteristic impedance is the impedance that a theoretically *infinite* length of this cable would present at its input end.

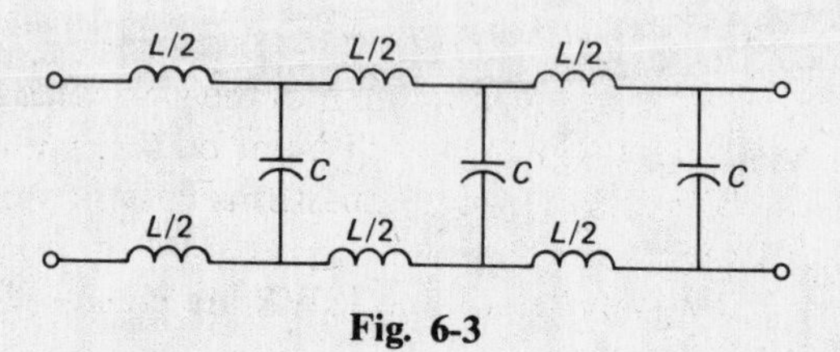

Fig. 6-3

Table 6-1

Type #	Description	Characteristic Impedance (Ω)
RG 8/U	Coaxial cable	52
RG11 A/U	Coaxial cable	75
214-056	Twin lead (commonly used for TV lead-in)	300
	Air-insulated parallel conductors with ceramic spacers	200–600

Since every section of cable has a capacitance and an inductance, an infinite length of cable can be considered as an infinite network of inductors and capacitors. See Fig. 6-3.

The relationship between capacitance and inductance per unit length and characteristic impedance is $Z_0 = \sqrt{L/C}$.

The characteristic impedance of a number of common cables used as transmission lines is shown in Table 6-1.

REFLECTED WAVES AND STANDING-WAVE RATIO (SWR)

Due to power and energy considerations at the load end of a short-circuited or open-circuited transmission line, an argument can be made for the existence of waves reflected from the load.

When a transmission line is terminated in any load other than a resistance having a magnitude equal to the characteristic impedance of the line, a reflected wave as well as an incident wave is present on the line. The summation of the incident and reflected wave at each point on the line gives rise to different rms voltage values at different points on the line.

A voltmeter placed at each point on the transmission line indicates an rms voltage at each point which varies from point to point on the line as shown in Fig. 6-4.

The ratio of the largest rms value to the smallest rms value of voltage on the line is called the *voltage standing-wave ratio* (VSWR). The largest rms and smallest rms values are measured at different points on the line separated by a distance equal to a quarter wavelength.

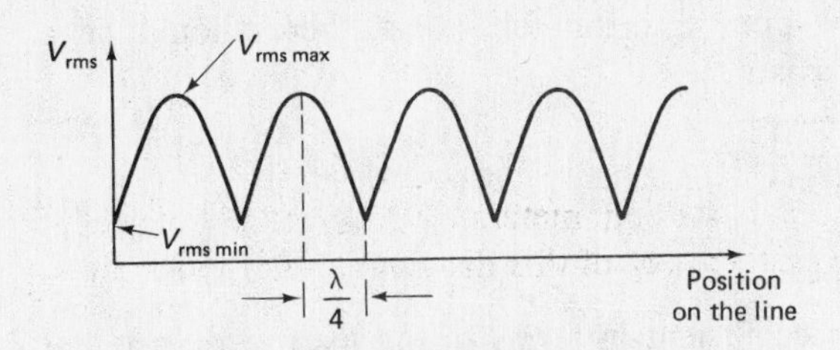

Fig. 6-4

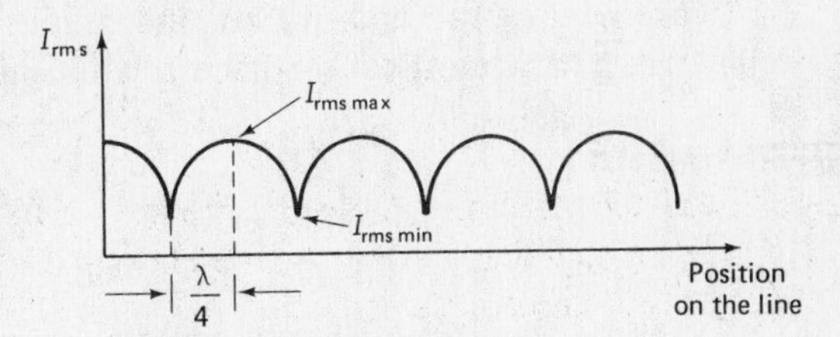

Fig. 6-5

Similarly, measurement of the rms value of current at each point on the transmission line gives rise to different values at each point on the line as shown in Fig. 6-5.

The ratio of the smallest rms current value to the largest is called the *current standing-wave ratio* (ISWR).

The VSWR and the ISWR are equal. Frequently SWR is used in place of the terms VSWR and ISWR.

STANDING-WAVE RATIO

The optimum condition for transmitting power to a load over a transmission line is one in which the maximum rms values of voltage and current are equal to the minimum rms values of voltage and current.

The SWR is an indication of how close or how far we are to the optimum condition for transmitting power to the load. The closer the SWR comes to 1 : 1, the closer the best or optimum condition is realized. In determining the SWR, a ratio is set up with the larger quantity taken first:

$$\text{SWR} = V_{\text{rms max}} : V_{\text{rms min}} = I_{\text{rms max}} : I_{\text{rms min}}$$

Laboratory and field investigation shows that

$$\text{SWR} = Z_L : Z_0$$

the SWR being a measure of the mismatch between load and line. As with all ratios, the SWR can also be represented as a fraction.

$$\text{SWR} = \frac{V_{\text{rms max}}}{V_{\text{rms min}}} = \frac{I_{\text{rms max}}}{I_{\text{rms min}}}$$

$$\text{SWR} = \frac{Z_L}{Z_0}$$

THE REFLECTION COEFFICIENT K_r

Another factor to be considered when dealing with mismatched load and transmission lines is the reflection coefficient K_r. K_r is defined as the reflected voltage divided by the incident voltage, or alternatively as reflected current divided by incident current.

Since in the optimum condition the reflected wave goes to zero, the optimum value for K_r is zero.

$$K_r = \frac{V_{\text{refl}}}{V_{\text{inc}}}$$

$$K_r = \frac{I_{\text{refl}}}{I_{\text{inc}}}$$

Since both SWR and K_r are indications of how poor a mismatch exists, there should be a relationship between them. Such a relationship does exist:

$$\text{SWR} = \frac{K_r + 1}{1 - K_r}$$

It should also be possible to express the reflection coefficient in terms of the load resistance and the characteristic impedance, their inequality to each other being the main cause of K_r being other than zero. This relationship also exists.

$$K_r = \left| \frac{Z_L - Z_0}{Z_0 + Z_L} \right|$$

REFLECTED POWER

Since the primary reason for launching a wave on a transmission line is to transfer power from a source to a load, we are concerned with how effectively power has been injected into the load. Since power is equal to the square of the voltage divided by the resistance of the load, and since the reflection coefficient is reflected voltage divided by incident voltage, we can write the following:

$$P = \frac{V^2}{R}$$

$$P_{\text{refl}} = \frac{V^2_{\text{refl}}}{R_L}$$

$$P_{\text{inc}} = \frac{V^2_{\text{inc}}}{R_L}$$

$$\frac{P_{\text{refl}}}{P_{\text{inc}}} = \frac{V_{\text{refl}}^2/R_L}{V_{\text{inc}}^2/R_L}$$

$$= \frac{V_{\text{refl}}^2}{V_{\text{inc}}^2}$$

$$K_r = \frac{V_{\text{refl}}}{V_{\text{inc}}}$$

$$\boxed{\frac{P_{\text{refl}}}{P_{\text{inc}}} = K_r^2}$$

VELOCITY FACTOR

Thus far in this chapter, the speed of electromagnetic waves on a transmission line has been approximated as being equal to the speed of electromagnetic waves in free space; 3×10^8 meters/s. This is not exactly true. In fact, the ratio of the speed of electromagnetic waves on a particular transmission line to that in free space is known as the velocity factor k. The velocity factor for most common transmission lines varies from a low of 0.55 for certain twisted pairs to 0.98 for small conductors widely spaced as open-wire lines.

QUARTER-WAVE MATCHING TRANSFORMERS

Quarter-wavelength sections of transmission line can be used to match a load to a transmission when the resistance of the load is not equal to the characteristic impedance of the transmission line. Since the quarter-wavelength section of line functions somewhat as a matching transformer, it is referred to as a quarter-wave matching transformer. The proper value of characteristic impedance for the quarter-wave transmission line section can be found from

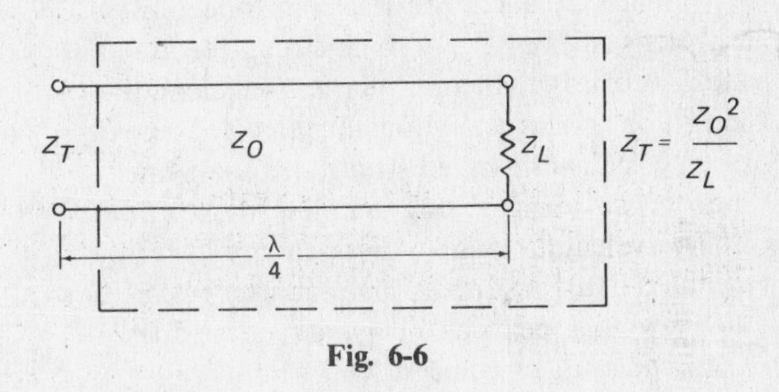

Fig. 6-6

$$Z_T = \frac{Z_0^2}{Z_L}$$

where Z_T is the impedance seen looking into the quarter-wave matching section when it is attached to the load Z_L, and Z_0 is the characteristic impedance of the quarter-wave matching section necessary to provide a match. See Fig. 6-6.

STUB MATCHING

Either because a load has a reactive component or because a resistive load is not matched to the characteristic impedance of the line, the impedance seen from the input end of a transmission line can have a reactive component. In order to eliminate the standing waves on such a transmission line, it is necessary to eliminate the reactive component of impedance as seen looking into the line from the input end. One way to do this is to place a short section of transmission line, either open-circuited or short-circuited, in parallel with the transmission line at a location close to the load end of the line. Such a short section of line is called a stub (see Fig. 6-7). Depending on the length of the short section of line and whether it is short-circuit-terminated or open-circuit-terminated,

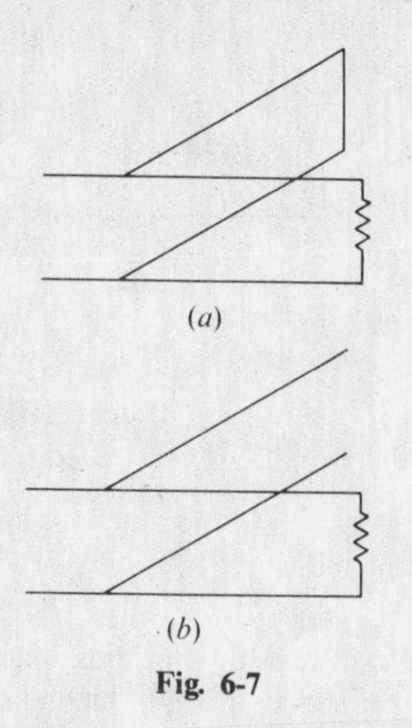

Fig. 6-7

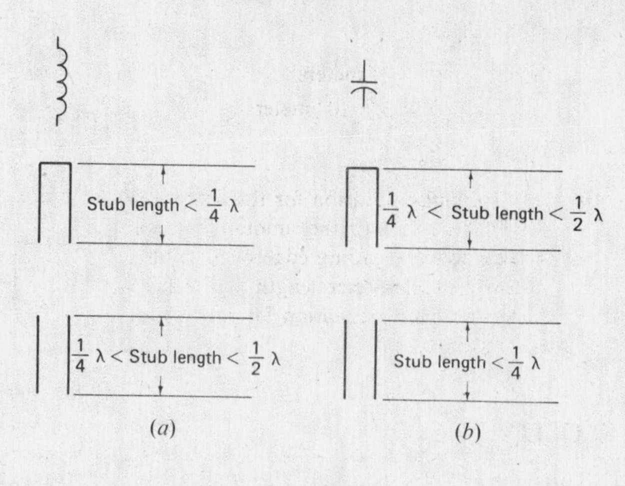

Fig. 6-8

it will appear as either a capacitive reactance or an inductance reactance (see Fig. 6-8). The proper choice of stub length and its installation at an appropriate position on the transmission line will cause the input end of the line to be looking into a resistive condition, and at the design frequency (or harmonic of the design frequency) no standing waves will exist from the input end of the line to the place on the line where the stub is located. If the frequency of the signal changes, then the matched condition provided by the stub will no longer prevail and an unmatched condition with standing waves will again exist.

Generally stubs are less than one half-wavelength long. A short-circuit-terminated stub less than one quarter-wavelength long appears as an inductance, as does an open-circuit-terminated stub more than one quarter-wavelength long but less than one half-wavelength long. An open-circuit-terminated stub less than one quarter-wavelength long appears as a capacitance, as does a short-circuit-terminated stub whose length is more than one quarter-wavelength but less than one half-wavelength (see Fig. 6-8).

Although the determination of the length and location of the stub usually requires the use of the Smith chart and/or complex algebraic solutions, a simplified approach is possible using available graphs such as the one shown in Fig. 6-9.

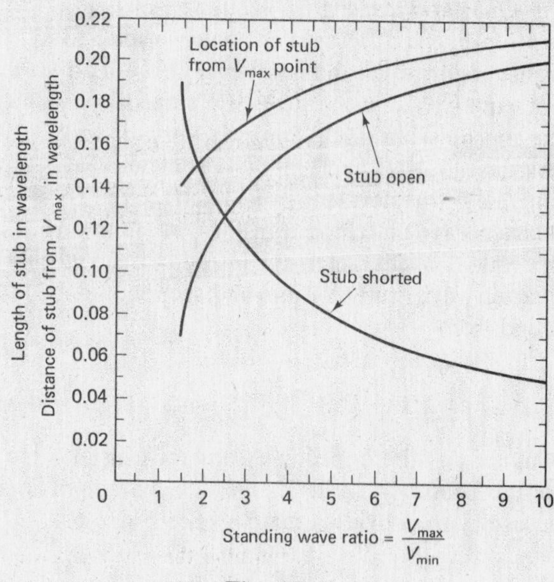

Fig. 6-9

Solved Problems

6.1 Determine the required pulse duration of a pulse so that when the pulse travels on a 25-meter line, the trailing edge occurs at the generator end of the line just as the leading edge reaches the load. Assume that the speed of the pulse on the line is the same as its free-space velocity (3×10^8 meters/s). See Fig. 6-10.

SOLUTION

Given: $d = 25$ meters
$c = 3 \times 10^8$ meters/s

Find: t_p

The pulse duration for this situation should be equal to the amount of time it takes for the leading edge of the pulse to travel the 25-meter length of the line. Using the classic equation for rate, time, and distance,

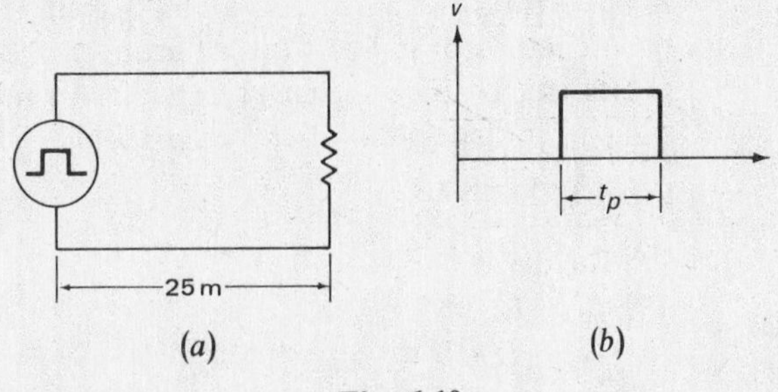

Fig. 6-10

$$\text{rate} \times \text{time} = \text{distance}$$
$$rt_p = d$$
$$(3 \times 10^8)t_p = 25$$
$$t_p = 8.33 \times 10^{-8}\ \text{s}$$
$$\boxed{t_p = 0.0833\ \mu\text{s}}$$

6.2 What is the wavelength of a 150-MHz sine wave? Make the determination first (*a*) in meters and then (*b*) in miles. Assume that the velocity of the wave on the line is the free-space velocity.

SOLUTION

Given: $f = 150$ MHz

Find: (*a*) λ in meters (*b*) λ in miles

(*a*) Assuming that the velocity of the wave on the transmission line is 3×10^8 meters/s, the wavelength in meters can be determined using

$$f\lambda = c$$
$$(150 \times 10^6)\lambda = 3 \times 10^8\ \text{meters/s}$$

Solving for λ,

$$\lambda = \frac{3 \times 10^8}{150 \times 10^6}$$
$$\boxed{\lambda = 2\ \text{meters}}$$

(*b*) In order to determine the wavelength in miles, either the above answer can be converted from meters to miles, or the formula $f\lambda = c$ can be used again, this time taking c to be 186 000 miles/s.

$$f\lambda = c$$
$$150 \times 10^6\lambda = 186\,000\ \text{miles/s}$$

Solving for λ,

$$\lambda = \frac{186\,000}{150 \times 10^6}$$
$$\boxed{\lambda = 0.001\,24\ \text{miles}}$$

6.3 A pulse train is transmitted along a transmission line which is 200 meters long. The pulse train consists of pulses with a duration of 30 ns each and separated by 45 ns. How many pulses can be on the line at any given time? Assume the speed of E/M waves to be the same as in free space. See Fig. 6-11.

SOLUTION

Given: $d = 200$ meters

$t_p = 30$ ns

$t_s = 45$ ns

Find: n

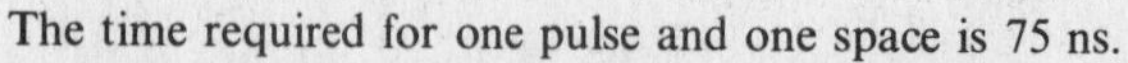

Fig. 6-11

$$\begin{aligned} t_T &= t_p + t_s \\ &= (30 \times 10^{-9}) + (45 \times 10^{-9}) \\ &= 75 \times 10^{-9} \end{aligned}$$

The time required for one pulse and one space is 75 ns.

Now determine the space available on the line for a total combination of one pulse and one space. Calculate the amount of time the pulse train remains on the line using

$$r \times t = d$$

Substitute numerical values and use 3×10^8 meters/s since the length of the line is given in meters.

$$3 \times 10^8 t = 200$$

Solving for t,

$$\begin{aligned} t &= \frac{200}{3 \times 10^8} \\ &= 66.667 \times 10^{-8} \\ &= 666.67 \times 10^{-9} \\ t_L &= 666.67 \text{ ns} \end{aligned}$$

Thus the length of the line represents 666.67 ns.

Now find the number of pulse and separation combinations that fit on the line:

$$\begin{aligned} t_L &= 666.67 \text{ ns} \\ t_T &= t_d + t_s \\ &= 30 \text{ ns} + 45 \text{ ns} \\ &= 75 \text{ ns} \\ n_T &= \frac{t_L}{t_T} = \frac{666.67 \times 10^{-9}}{75 \times 10^{-9}} \\ &= 8.8889 \end{aligned}$$

Thus eight full pulses and space combinations fit on the line with some room left over. The next question is whether the remaining space is enough for another full pulse.

The eight combinations occupy

$$8 \times 75 \text{ ns} \quad \text{or} \quad 600 \text{ ns}$$

with

$$666.67 - 600 = 66.67 \text{ ns}$$

left over.

Since each pulse duration is 30 ns, there is room for 1 more whole pulse in addition to the 8 pulse and space combinations.

Thus there are a maximum of $8 + 1 = 9$ pulses on the line. So,

$$\boxed{n = 9}$$

6.4 How many 600-kHz waves can be on a 5-mile transmission line simultaneously?

SOLUTION

Given: $f = 600$ kHz
$d = 5$ miles

Find: n

First determine the wavelength of the 600-kHz signal using

$$f\lambda = c$$

Use 186 000 miles per hour as the speed of light since the line length is given in miles and no figure is given for the velocity factor.

Substituting numerical values,

$$600 \times 10^3 \lambda = 186\,000$$
$$\lambda = \frac{186\,000}{600 \times 10^3}$$
$$= 0.31 \text{ miles}$$

Knowing the wavelength of the signal and the length of the line, the number of cycles on the line can be found from

$$n = \frac{d}{\lambda}$$
$$= \frac{5}{0.31}$$
$$\boxed{n = 16.13}$$

6.5 A sine wave having a frequency of 75 MHz is launched on a transmission line.

(*a*) How long does it take from the time that the instantaneous voltage is zero before a peak occurs at the launch point?

(*b*) How far along the transmission line has the leading edge of the wavefront progressed in this amount of time?

Assume the speed of electromagnetic waves on this line to be the same as in free space (3×10^8 meters/s or 186 000 miles/s).

SOLUTION

Given: $f = 75$ MHz

Find: (*a*) $T/4$ (*b*) d

(*a*) It will take one quarter period for the wave to go from zero to a peak. See Fig. 6-12. First determine the period and then determine one quarter of the period.

Determine the period T using the classical equation

$$T = \frac{1}{f}$$

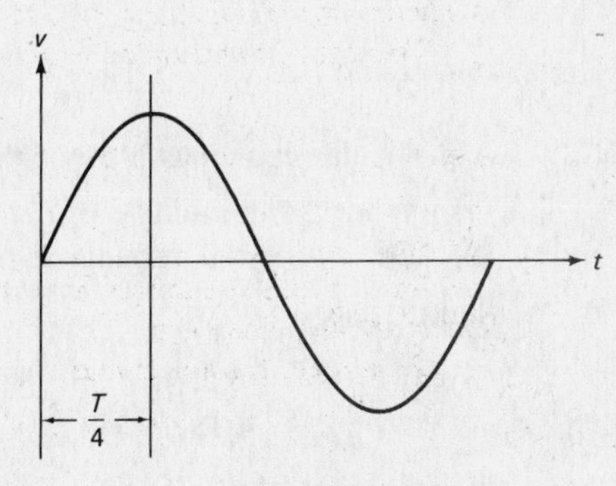

Fig. 6-12

Substituting numerical values,

$$T = \frac{1}{75 \times 10^6}$$
$$= 13.3 \times 15^{-9} \text{ s}$$

Thus the period is 13.3 ns.

One quarter of the full wavelength will provide the distance sought:

$$\frac{T}{4} = \frac{13.3 \times 10^{-9}}{4}$$
$$= 3.33 \times 10^{-9}$$

or

$$\boxed{\frac{T}{4} = 3.33 \text{ ns}}$$

(*b*) The distance that the leading edge has traveled in this amount of time can be calculated using the classical equation

$$d = r \times t$$

Substituting numerically,

$$d = (3 \times 10^8)(3.33 \times 10^{-9})$$

$$\boxed{d = 1.00 \text{ meter}}$$

6.6 Determine the characteristic impedance of a transmission line which has a capacitance of 35 pF/ft and an inductance of 0.25 μH/ft.

SOLUTION

Given: $c = 35$ pF/ft
$L = 0.25$ μH/ft

Find: Z_0

Using the equation relating Z_0, L, and C,

$$Z_0 = \sqrt{L/C}$$

Substituting numerical values,

$$Z_0 = \sqrt{\frac{0.25 \times 10^{-6}}{35 \times 10^{-12}}}$$

$$\boxed{Z_0 = 84.5\ \Omega}$$

6.7 A particular cable has a capacitance of 40 pF/ft and a characteristic impedance of 70 Ω.

(*a*) What is the inductance per foot of this cable?
(*b*) Determine the impedance of an infinitely long section of such cable.

SOLUTION

Given: $C = 40$ pF/ft
$Z_0 = 70\ \Omega$

Find: (*a*) L (*b*) Z

(*a*) Using the formula relating characteristic impedance, capacitance, and inductance,

$$Z_0 = \sqrt{L/C}$$

Substituting numerical values,

$$70 = \sqrt{\frac{L}{40 \times 10^{-12}}}$$

Now, in order to get L out from under the radical sign, square both sides of the equation:

$$70^2 = \left[\sqrt{\frac{L}{40 \times 10^{-12}}}\right]^2$$

which results in

$$70^2 = \frac{L}{40 \times 10^{-12}}$$

Multiplying both sides of the equation by (40×10^{-12}), L can be obtained by itself on one side of the equals sign:

$$(70)^2(40 \times 10^{-12}) = L$$

Rearranging,

$$L = 1.96 \times 10^{-7}$$
$$= 0.196 \times 10^{-6}$$

$$\boxed{L = 0.196\ \mu\text{H}}$$

(*b*) The characteristic impedance of a transmission line is the impedance that an infinite length of the line would present to a power supply at the input end of the line. So, in this case,

$$\boxed{Z_\infty = Z_0 = 70\ \Omega}$$

6.8 Voltage and current readings are taken on a transmission line at different points. The maximum voltage reading is 60 $V_{\text{rms max}}$, and the minimum voltage reading is 20 $V_{\text{rms max}}$.

(*a*) Calculate the VSWR on this line.
(*b*) What is the ISWR on this line?
(*c*) If the maximum current reading on the line is 2.5 A, what would the lowest current reading be?

SOLUTION

Given: $V_{\text{rms max}} = 60$ V
$V_{\text{rms min}} = 20$ V
$I_{\text{rms max}} = 2.5$ A

Find: (*a*) VSWR (*b*) ISWR (*c*) $I_{\text{rms min}}$

(*a*) By definition,

$$\text{VSWR} = V_{\text{rms max}} : V_{\text{rms min}}$$
$$= 60 : 20$$

$$\boxed{\text{VSWR} = 3 : 1}$$

(*b*) Since ISWR = VSWR,

$$\boxed{\text{ISWR} = 3:1}$$

(*c*)

$$\text{ISWR} = I_{\text{rms max}} : I_{\text{rms min}}$$
$$3:1 = 2.5 : I_{\text{rms min}}$$

Converting to fractions,

$$\frac{3}{1} = \frac{2.5}{I_{\text{rms min}}}$$
$$I_{\text{rms min}} = \frac{2.5}{3}$$

Thus,

$$\boxed{I_{\text{rms min}} = 0.833 \text{ A}}$$

6.9 A transmission line having a characteristic impedance of 75 Ω is delivering power to a 150-Ω load.

(*a*) Calculate the SWR on this line.
(*b*) Determine the minimum voltage reading on this line if the maximum voltage is 25 V.

SOLUTION

Given: $Z_0 = 75\ \Omega$
$Z_L = 150\ \Omega$
$V_{\text{rms max}} = 25$ V

Find: (*a*) SWR (*b*) $V_{\text{rms min}}$

(*a*) Since SWR, VSWR, and ISWR are all equal, determination of one provides all three.
From the equation relating SWR, Z_L, and Z_0,

$$\text{SWR} = \frac{Z_L}{Z_0}$$
$$= \frac{150}{75}$$

$$\boxed{\text{SWR} = 2:1}$$

(*b*)

$$\text{VSWR} = \text{SWR} = \frac{V_{\text{rms max}}}{V_{\text{rms min}}}$$
$$2 = \frac{25}{V_{\text{rms min}}}$$
$$V_{\text{rms min}} = \frac{25}{2}$$

$$\boxed{V_{\text{rms min}} = 12.5 \text{ V}}$$

6.10 A 50-Ω load is being fed from a 72-Ω transmission line. See Fig. 6-13.

(*a*) What is the standing-wave ratio resulting from this mismatch?

(*b*) Determine the reflection coefficient resulting from this mismatch.

(*c*) What percentage of the indicent power is reflected from the load?

(*d*) What percentage of the incident power is absorbed by the load?

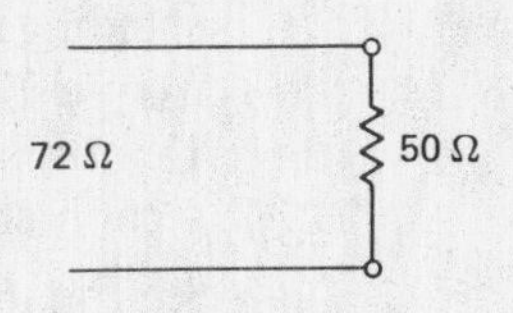

Fig. 6-13

SOLUTION

Given: $Z_0 = 72\ \Omega$
$Z_L = 50\ \Omega$

Find: (*a*) SWR (*b*) K_r (*c*) $\dfrac{P_{\text{refl}}}{P_{\text{inc}}} \times 100$ (*d*) $\dfrac{P_{\text{abs}}}{P_{\text{inc}}} \times 100$

(*a*) Setting up a ratio with the larger quantity taken first, the SWR can be found from

$$\begin{aligned} \text{SWR} &= Z_0 : Z_L \\ &= 72 : 50 \\ &= \frac{72}{50} : 1 \end{aligned}$$

$$\boxed{\text{SWR} = 1.44 : 1}$$

(*b*) The reflection coefficient is related to Z_L and Z_0 by

$$\begin{aligned} K_r &= \left| \frac{Z_L - Z_0}{Z_0 + Z_L} \right| \\ &= \frac{72 - 50}{72 + 50} \\ &= \frac{22}{122} \end{aligned}$$

$$\boxed{K_r = 0.180}$$

(*c*) Since K_r is a ratio of voltages at the load, K_r^2 is a ratio of powers:

$$\begin{aligned} \frac{P_{\text{refl}}}{P_{\text{inc}}} &= K_r^2 \\ &= (0.180)^2 = 0.0324 \end{aligned}$$

Converting from a decimal to a percentage by multiplying by 100,

$$\% P_{\text{refl}} = 0.0324 \times 100$$

$$\boxed{\% P_{\text{refl}} = 3.24\%}$$

(*d*) Knowing the percent of power reflected, the power absorbed is that left over from 100%.

$$\% P_{\text{abs}} = 100 - 3.24$$

$$\boxed{\% P_{\text{abs}} = 96.8\%}$$

6.11 (*a*) Determine the required length of a quarter-wave matching section which will eliminate standing waves and thereby provide a matched condition for a 300-Ω resistive load fed from a 72-Ω transmission line. This condition is to exist for a frequency of 100 MHz.

(*b*) Determine the characteristic impedance of the transmission line from which the matching section should be cut. Assume a velocity factor of 1.0.

See Fig. 6-14.

SOLUTION

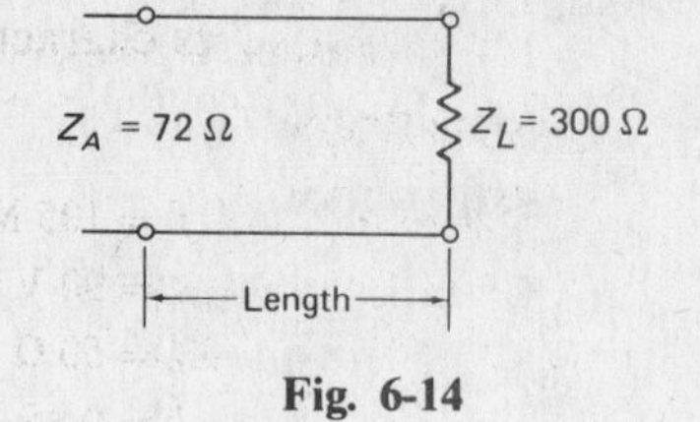

Fig. 6-14

Given: $Z_L = 300\ \Omega$

$Z_A = 72\ \Omega$

$f = 100$ MHz

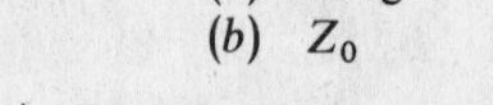

Find: (*a*) Length

(*b*) Z_0

(*a*) In order to determine the length of a quarter-wave matching section, it is first necessary to determine the wavelength of the 100-MHz signal:

$$f\lambda = C$$

$$\lambda = \frac{C}{f} = \frac{3 \times 10^8}{100 \times 10^6} = 3 \text{ meters}$$

$$\text{Length} = 0.25\ \lambda = 0.25\ (3)$$

$$\boxed{\text{Length} = 0.75 \text{ meter}}$$

(*b*) The characteristic impedance of the required matching section is related to the load impedance and the original line by

$$Z_0 = \sqrt{Z_A Z_L}$$

Solving,

$$Z_0 = \sqrt{72(300)} = \sqrt{21\,600}$$

$$\boxed{Z_0 = 146.97\ \Omega}$$

6.12 A 105-MHz, 90-V peak signal is incident on a 50-Ω transmission line. The velocity factor k of this line is 0.85. The line is 125 meters long and is terminated in a 300-Ω load. See Fig. 6-15.

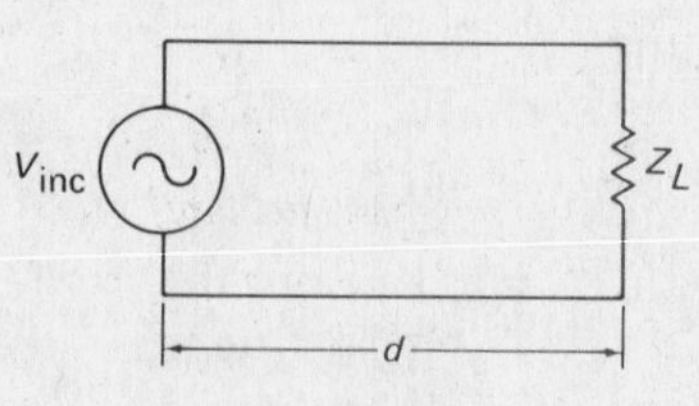

Fig. 6-15

(*a*) Find the wavelength of the signal on the line.
(*b*) Determine the length of the line in wavelengths.
(*c*) What is the SWR for this situation?
(*d*) Find the reflection coefficient.
(*e*) Calculate the peak value of the reflected voltage wave.

(*f*) What percent of the incident power is returned as reflected power?
(*g*) Find the peak values of I_{inc} and I_{refl}.
(*h*) Determine the peak value of the voltage standing wave at the voltage antinodes.
(*i*) Determine the peak value of the current standing wave at the current antinodes.
(*j*) Determine the peak value of the voltage standing wave at the voltage nodes.
(*k*) Determine the peak value of the current standing wave at the current nodes.
(*l*) If a quarter-wavelength matching section is to be used to correct for a mismatch, what must be its characteristic impedance?

SOLUTION

Given:

$$f = 105 \text{ MHz}$$
$$V_{\text{inc peak}} = 90 \text{ V}$$
$$Z_0 = 50\ \Omega$$
$$k = 0.85$$
$$d = 125 \text{ meters}$$
$$Z_L = 300\ \Omega$$

Find:

(*a*) Wavelength
(*b*) d in wavelengths
(*c*) SWR
(*d*) K_r
(*e*) $V_{\text{refl peak}}$
(*f*) Percent power reflected
(*g*) $I_{\text{inc peak}}$, $I_{\text{refl peak}}$
(*h*) $V_{\text{peak total}}$ at antinodes
(*i*) $I_{\text{peak total}}$ at antinodes
(*j*) $V_{\text{peak total}}$ at nodes
(*k*) $I_{\text{peak total}}$ at nodes
(*l*) Z_0

(*a*) Taking into consideration the velocity factor k, the wavelength λ can be found from

$$f\lambda = kC$$

Substituting,

$$105 \times 10^6 \lambda = 0.85 \times 3 \times 10^8$$
$$\lambda = \frac{0.85 \times 3 \times 10^8}{105 \times 10^6}$$

$$\boxed{\lambda = 2.43 \text{ meters}}$$

(*b*) Dividing the length of the line by the wavelength of the signal provides us with the length of the line in wavelengths:

$$\text{Wavelengths} = \frac{d}{\lambda} = \frac{125}{2.43}$$

Thus,

$$\boxed{d = 51.44 \text{ wavelengths}}$$

(*c*) The standing-wave ratio is determined from Z_L and Z_0 by

$$\text{SWR} = Z_L : Z_0 = \frac{Z_L}{Z_0} : 1$$

Substituting,

$$\text{SWR} = \frac{300}{50} : 1$$

$$\boxed{\text{SWR} = 6 : 1}$$

(*d*) The reflection coefficient can be found from

$$K_r = \left| \frac{Z_L - Z_0}{Z_L + Z_0} \right|$$

Substituting,

$$K_r = \frac{250}{350}$$

$$\boxed{K_r = 0.715}$$

(*e*) Knowing the reflection coefficient and incident voltage, we can determine the peak value of the reflected voltage:

$$K_r = \frac{V_{\text{refl}}}{V_{\text{inc}}}$$

$$0.715 = \frac{V_{\text{refl peak}}}{90}$$

$$\boxed{V_{\text{refl peak}} = 64.35 \text{ V}}$$

(*f*) Power relationships are equal to the square of voltage relationships for the same load $(P = V^2/R)$. Thus,

$$\frac{P_{\text{refl}}}{P_{\text{inc}}} = K_r^2 = (0.715)^2$$

So,

$$\% \text{ power reflected} = \frac{P_{\text{refl}}}{P_{\text{inc}}} \times 100 = K_r^2 \times 100$$

$$= (0.715)^2 \times 100$$

$$\boxed{\% \text{ power reflected} = 51\%}$$

(*g*) Applying Ohm's law at any point on the transmission line,

$$\frac{V_{\text{inc}}}{I_{\text{inc}}} = Z_0$$

$$I_{\text{inc peak}} = \frac{90}{50}$$

$$\boxed{I_{\text{inc peak}} = 1.8 \text{ A}}$$

Similarly,

$$\frac{V_{\text{refl}}}{I_{\text{refl}}} = Z_0$$

$$I_{\text{refl peak}} = \frac{64.35}{50}$$

$$\boxed{I_{\text{refl peak}} = 1.287 \text{ A}}$$

(*h*) The antinodes are those points on the transmission line at which the incident and reflected waves totally reinforce each other:

$$V_{\text{peak total max}} = V_{\text{inc peak}} + V_{\text{refl peak}}$$
$$= 90 + 64.35$$

$$\boxed{V_{\text{peak total max}} = 154.35 \text{ V}}$$

(*i*)

$$I_{\text{peak total max}} = I_{\text{inc peak}} + I_{\text{refl peak}}$$
$$= 1.8 + 1.287$$

$$\boxed{I_{\text{peak total max}} = 3.087 \text{ A}}$$

(*j*) The nodes are those points on the transmission line at which the incident and reflected waves are in maximum opposition to each other.

$$V_{\text{peak total min}} = V_{\text{inc peak}} - V_{\text{refl peak}}$$
$$= 90 - 64.35$$

$$\boxed{V_{\text{peak total min}} = 25.65 \text{ V}}$$

(*k*)

$$I_{\text{peak total min}} = I_{\text{inc peak}} - I_{\text{refl peak}}$$
$$= 1.8 - 1.287$$

$$\boxed{I_{\text{peak total min}} = 0.513 \text{ A}}$$

(*l*) The characteristic impedance of the matching section can be found from

$$Z_0 = \sqrt{Z_A Z_L}$$

Substituting,

$$Z_0 = \sqrt{50(300)}$$
$$= \sqrt{15\,000}$$

$$\boxed{Z_0 = 122.47 \ \Omega}$$

6.13 A short-circuited stub is required for use in a situation in which there is an SWR of 4 : 1 on a transmission line. Determine the length of the stub and its appropriate location on the line. The frequency at which standing waves are to be eliminated is 200 MHz.

SOLUTION

Given: SWR = 4 : 1
$f = 200$ MHz

Find: L_s, d_s

Using Fig. 6-9 and entering the graph at SWR = 4, it is seen that the length of the shorted stub must equal 0.09λ and be placed 0.181λ from the voltage antinode. The wavelength of the 100-MHz signal is

$$f\lambda = 3 \times 10^8$$
$$200 \times 10^6 \lambda = 3 \times 10^8$$
$$\lambda = \frac{3 \times 10^8}{200 \times 10^6}$$
$$\lambda = 1.5 \text{ meters}$$

Stub length is thus

$$L_s = 0.09(1.5)$$

$$\boxed{L_s = 0.135 \text{ meter}}$$

The distance from the voltage antinode is

$$d_s = 0.181(1.5)$$

$$\boxed{d_s = 0.2715 \text{ meter}}$$

Supplementary Problems

6.14 A pulse has a duration of 20 μs. How far down the transmission line does the leading edge get by the time the trailing edge appears at the generator end of the line? *Ans.* 6000 meters

6.15 Determine the pulse duration of a pulse such that its trailing edge occurs at the generator just as its leading edge gets halfway down a 10-meter transmission line. Assume the speed of E/M waves on the line to be the same as in free space. *Ans.* 16.67 ns

6.16 A transmission line is 80 meters long. The line is being fed a pulse train consisting of pulses each having a duration of 15 ns and spaced 25 ns apart. How many pulses can be on the line at any given time? Assume that the speed of E/M waves on the line is the same as in free space.
Ans. 7 pulses

6.17 A 3-mile transmission line is fed a pulse train consisting of pulses each having a duration of 40 ns and separated from each other by 70 ns. How many pulses can be on the line at any given time?
Ans. 147

6.18 How many 1-MHz waves can be on a 4-mile transmission line simultaneously? *Ans.* 21.5

6.19 On a transmission line 4.5 cycles of a 2.5-MHz signal appear simultaneously. How long is the line in miles and in meters? *Ans.* 0.3348 mile, 540 meters

6.20 A transmission line is 500 meters long. How many pulses of a pulse train are on the line simultaneously? The pulse duration is 150 ns while rest time between pulses is 250 ns. *Ans.* 4+

6.21 Calculate the wavelength in free space of the following electromagnetic signals in meters, in miles, and in feet.

(*a*) 20 Hz (*c*) 2000 Hz (*e*) 200 000 Hz (*g*) 20 MHz
(*b*) 200 Hz (*d*) 20 000 Hz (*f*) 2 MHz (*h*) 200 MHz

Ans. Meters: (*a*) 15 000 k (*c*) 150 k (*e*) 1.5 k (*g*) 15
(*b*) 1500 k (*d*) 15 k (*f*) 150 (*h*) 1.5
Miles: (*a*) 9300 (*c*) 93 (*e*) 0.93 (*g*) 0.0093
(*b*) 930 (*d*) 9.3 (*f*) 0.093 (*h*) 0.000 93
Feet: (*a*) 49 104 000 (*c*) 491 040 (*e*) 4910 (*g*) 49.1
(*b*) 4 910 400 (*d*) 49 104 (*f*) 491 (*h*) 4.91

6.22 A cable has a capacitance of 10 pF/ft and an inductance of 0.03 μH/ft. What is the characteristic impedance of the cable? *Ans.* 54.77 Ω

6.23 Determine the characteristic impedance of a transmission line which has a capacitance of 12 pF/ft and an inductance of 0.015 μH/ft.
Ans. 35.36 Ω

6.24 Calculate the characteristic impedance of a cable having a capacitance of 25 pF/ft and an inductance of 0.18 μH/ft. *Ans.* 84.85 Ω

6.25 Find the characteristic impedance of a transmission line that has a capacitance of 20 pF/ft and an inductance of 0.22 μH/ft. *Ans.* 104.88 Ω

6.26 Find the capacitance per foot of a transmission line having a characteristic impedance of 80 Ω and an inductance of 0.30 μH/ft. *Ans.* 46.875 pF

6.27 Calculate the capacitance per foot of a cable that has a characteristic impedance of 75 Ω and an inductance of 0.19 μH/ft. *Ans.* 33.778 pF

6.28 A cable has a capacitance of 40 pF/ft and a characteristic impedance of 300 Ω. What is the inductance per foot of this cable? *Ans.* 3.6 μH

6.29 Find the inductance per foot of a transmission line that has a characteristic impedance of 200 Ω and a capacitance of 35 pF/ft. *Ans.* 1.4 μH

6.30 Determine the standing-wave ratio on a transmission line on which the maximum rms voltage is 90 V and the minimum rms voltage is 25 V. *Ans.* 3.6 : 1

6.31 What is the SWR on a transmission line which has a maximum rms voltage of 118 V and a minimum rms voltage of 50 V? *Ans.* 2.36 : 1

6.32 A 1.5-MHz sine wave is launched on a transmission line. How much time does it take from the instant that the instantaneous voltage is zero until a peak occurs at the launch point? How far along the transmission line has the leading edge of the wavefront progressed in this amount of time?

Assume the speed of electromagnetic waves on this line to be the same as in free space (3×10^8 meters/s or 186 000 miles/s). *Ans.* 166.67 ns, 50 meters

6.33 A 40-MHz sine wave is transmitted on a transmission line. How much time elapses between the occurrence of instantaneous zero values at the launch point? How far down the transmission line has the first incident instantaneous zero value progressed when the second zero occurs at the launch point? *Ans.* 12.5 ns, 3.75 meters

6.34 Find the SWR on a transmission line having a maximum rms current of 1.75 A and a minimum of 0.85 A at different points on the line. *Ans.* 2.059 : 1

6.35 Voltage and current readings are taken at many different points on a transmission line. The maximum voltage reading is 140 V rms and a minimum voltage reading of 65 V rms. The maximum current reading on the line is 4.8 A.

(*a*) Calculate the VSWR on this line.
(*b*) What is the ISWR on this line?
(*c*) Determine the lowest current reading on the line.

Ans. (*a*) 2.1538 : 1, (*b*) 2.1538 : 1, (*c*) 2.229 A

6.36 The maximum rms current reading anywhere along a particular transmission line is 8.4 A while the minimum rms current reading anywhere on the line is 2.8 A.

(*a*) What is the ISWR on this line?
(*b*) Determine the VSWR on this line.
(*c*) If the maximum rms voltage on this line is 178 V, what is the minimum rms voltage on the line?

Ans. (*a*) 3 : 1, (*b*) 3 : 1, (*c*) 59.33 V

6.37 A 250-Ω transmission line carries a signal from a source to a 40-Ω load. Determine the standing-wave ratio on the line. *Ans.* 6.25 : 1

6.38 A 50-Ω transmission line delivers energy to a 300-Ω source. What is the SWR on the line? *Ans.* 6 : 1

6.39 A 300-Ω transmission line delivers energy to a 65-Ω load. Calculate the SWR on the line. *Ans.* 4.6 : 1

6.40 A 72-Ω transmission line has a standing-wave ratio of 8 : 1 due to a mismatch between the line and the resistive load. What is the resistance of the load? *Ans.* 9 Ω or 576 Ω

6.41 A 4.3 : 1 standing-wave ratio appears on a transmission line due to a mismatch between the line impedance and the impedance of a resistive load. Determine the resistance of the load if the characteristic impedance of the line is 140 Ω. *Ans.* 32.56 Ω or 602 Ω

6.42 A standing-wave ratio of 2 : 1 results when a certain transmission line is connected to a 75-Ω resistive load. What is the characteristic impedance of the transmission line? *Ans.* 150 Ω or 37.5 Ω

6.43 An SWR of 3 : 1 results when a transmission line is terminated in a 90-Ω load. Determine the characteristic impedance of the transmission line. *Ans.* 30 Ω or 270 Ω

6.44 What is the characteristic impedance of a transmission line if terminating it in a 300-Ω load results in an SWR of 2 : 1? *Ans.* 150 Ω or 600 Ω

6.45 Calculate the reflection coefficient on a transmission line having an SWR of 3 : 1. *Ans.* .5

6.46 Determine the reflection coefficient on a transmission line having an SWR of 2.5 : 1. *Ans.* .429

6.47 (*a*) Determine the SWR and the reflection coefficient for a transmission line whose characteristic impedance is 50 Ω and which is terminated in a 90-Ω resistive load.
(*b*) What percentage of the incident power is reflected and what percentage is absorbed by the load?
Ans. (*a*) 1.8 : 1, .286; (*b*) 8.16%, 91.84%

6.48 (*a*) What SWR and reflection coefficient result when a 300-Ω transmission line is used to feed a 70-Ω load?
(*b*) What percentage of the incident power is reflected by the load and what percentage is absorbed by the load?
Ans. (*a*) 4.286 : 1, .622; (*b*) 38.6%, 61.4%

6.49 What is the necessary length and characteristic impedance of a cable to be used as a quarter-wave matching transformer which is to eliminate the standing waves due to the mismatch between a 50-Ω transmission line and a 300-Ω load? The match is desired at 60 MHz. Assume a velocity factor of 1.0. *Ans.* 1.25 meters, 122.47 Ω

6.50 Calculate the length and characteristic impedance of a transmission line which is to act as a quarter-wave matching transformer between a 180 Ω transmission line and a 400-Ω load. The matching action is to be effective at 600 kHz. Assume a velocity factor of 1.0. *Ans.* 125 meters, 268.33 Ω

6.51 Find the required length and characteristic impedance of a quarter-wave matching transformer to be used to eliminate standing waves due to a mismatch between a 135-Ω transmission line and a 40-Ω resistive load at 144 MHz. Assume a velocity factor of 1.0. *Ans.* 0.521 meter, 73.48 Ω

6.52 A 50-Ω cable that is 0.7 meter long is being used to match a 300-Ω transmission line to a load. Determine the resistance of the load and the frequency at which the match is supposed to occur. Assume a velocity factor of 1.0. *Ans.* 8.33 Ω, 107 MHz

6.53 Determine the appropriate length for a quarter-wave matching section to be effective at 55 MHz if the velocity factor on this line is 0.85. *Ans.* 1.159 meters

6.54 For what frequency is a 2.3-meter quarter-wave matching section cut if the velocity factor on this line is 0.92? *Ans.* 30 MHz

6.55 A 250-MHz 40-V peak signal is incident on a 72-Ω transmission line. The velocity factor for this line is 0.91. The line is 250 meters long and is terminated in a 200-Ω load.

(*a*) Find the wavelength of the signal on the line.
(*b*) Determine the length of the line in wavelengths.
(*c*) What is the SWR for this situation?
(*d*) Find the reflection coefficient.
(*e*) Calculate the peak value of the reflected voltage wave.
(*f*) What percent of the incident power is returned as reflected power?
(*g*) Find the peak values of I_{inc} and I_{refl}.
(*h*) Determine the peak value of the voltage standing wave at the voltage antinodes.
(*i*) Determine the peak value of the current standing wave at the current antinodes.
(*j*) Determine the peak value of the voltage standing wave at the voltage nodes.
(*k*) Determine the peak value of the current standing wave at the current nodes.
(*l*) If a quarter-wavelength matching section is to be used to correct for a mismatch, what must its characteristic impedance be?

Ans. (*a*) 1.092 meters; (*b*) 228.94 wavelengths; (*c*) 2.778 : 1; (*d*) 0.4706; (*e*) 18.82 V; (*f*) 22.146%; (*g*) 0.5556 A, 0.2614 A; (*h*) 58.82 V; (*i*) 0.817 A; (*j*) 21.18 V; (*k*) 0.2942 A; (*l*) 120 Ω

6.56 A 112.6-MHz 75-V peak signal is incident on a 150-Ω transmission line. The velocity factor for this line is 0.89. This line is 175 meters long and is terminated in a 70-Ω load.

(*a*) Find the wavelength of the signal on the line.
(*b*) Determine the length of the line in wavelengths.
(*c*) What is the SWR for this situation?
(*d*) Find the reflection coefficient.
(*e*) Calculate the peak value of the reflected voltage wave.
(*f*) What percent of the incident power is returned as reflected power?
(*g*) Find the peak values of I_{inc} and I_{refl}.
(*h*) Determine the peak value of the voltage standing wave at the voltage antinodes.
(*i*) Determine the peak value of the current standing wave at the current antinodes.
(*j*) Determine the peak value of the voltage standing wave at the voltage nodes.
(*k*) Determine the peak value of the current standing wave at the current nodes.
(*l*) If a quarter-wavelength matching section is to be used to correct for a mismatch, what must its characteristic impedance be?

Ans. (*a*) 2.37 meters; (*b*) 73.84 wavelengths; (*c*) 2.143 : 1; (*d*) 3636; (*e*) 27.27 V; (*f*) 13.22%; (*g*) 0.5 A, 0.1818 A; (*h*) 102.27 V; (*i*) 0.6818 A; (*j*) 47.73 V; (*k*) 0.3182 A; (*l*) 102.47 Ω

6.57 Determine the length of a short-circuited stub and its location in order to eliminate an SWR of 3 : 1 on a transmission line handling a 75-MHz signal. Assume a velocity factor of 1.0.
Ans. 0.448 meter, 0.68 meter

6.58 What is the proper length and location of a short-circuited stub in order for it to eliminate a 2.5 : 1 SWR at a frequency of 144 MHz? Assume a velocity factor of 0.88 for the line.
Ans. 0.238 meter, 0.2988 meter

Chapter 7

Antennas

INTRODUCTION

An antenna is a device whose function is to radiate electromagnetic energy and/or intercept electromagnetic radiation. A transmitting antenna can be used for reception and vice versa. In two-way communications, the same antenna is used for both transmission and reception.

RADIATION PATTERNS

Antennas do not necessarily perform equally well in all directions. A polar diagram which indicates how well an antenna transmits or receives in different directions is called a radiation pattern.

Figure 7-1 is a radiation pattern for an antenna known as a *Marconi antenna.*

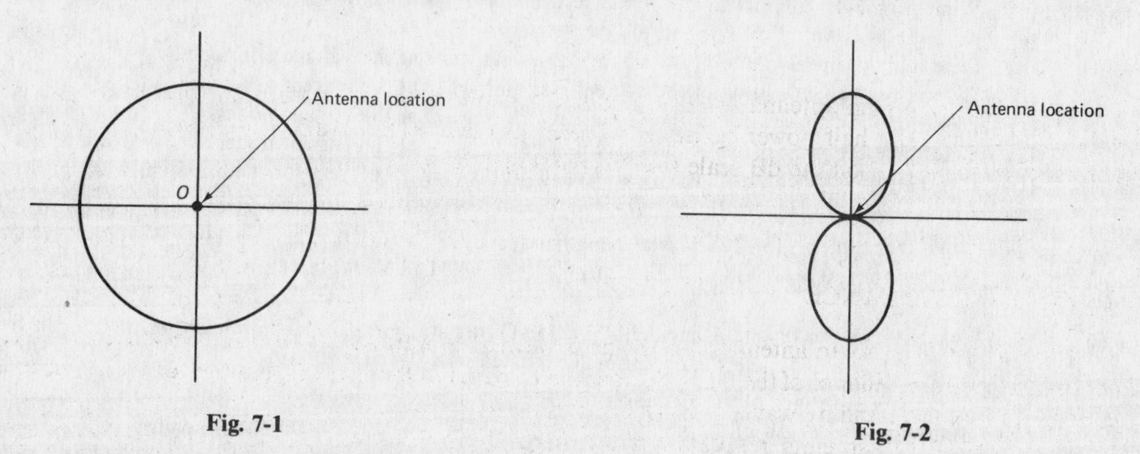

Fig. 7-1

Fig. 7-2

Figure 7-2 is a radiation pattern for an antenna known as a *Hertz antenna.* The distance from the location of the antenna to a point on the radiation pattern indicates the relative strength of the radiation in the direction determined by these two points.

Once a radiation-pattern diagram exists, as in Fig. 7-3, to determine the relative strength of the signal at point A, a line is drawn from the origin, point O, to point A. The intercept of this line with the radiation pattern determines the termination of the vector. When an antenna is being used for receiving purposes, the radiation pattern becomes a *reception pattern.* The long section of the lobes of the pattern indicates the best direction for reception.

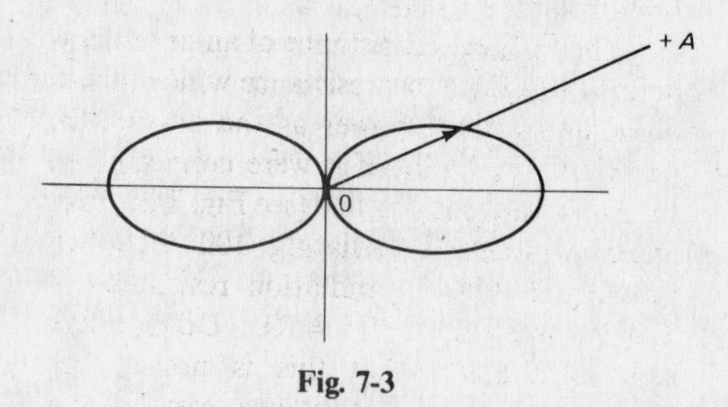

Fig. 7-3

BEAM WIDTH

It is frequently necessary to have a quick means of comparing the directivity of antennas without going through a point-by-point comparison of their radiation patterns. The beam width is such a quick description. The beam width of an antenna is the angle within which the power radiated is above one-half of what it is in the most preferential direction, or it can be said that the beam width is the angle within which the voltage developed by a receiving antenna remains within 70.7% of the

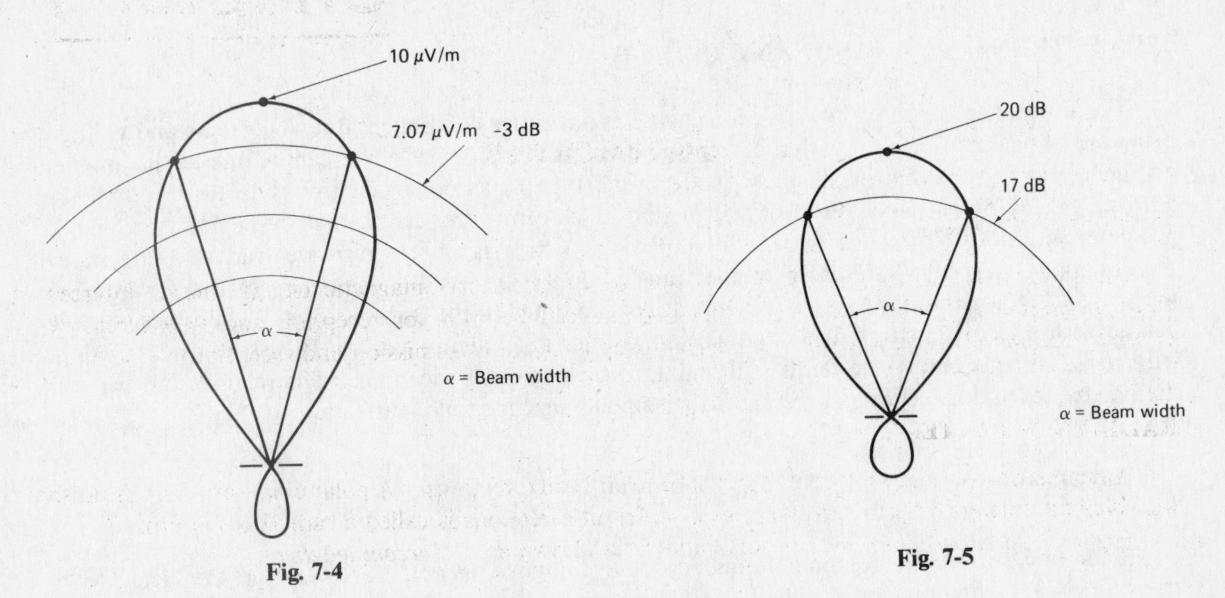

Fig. 7-4

Fig. 7-5

voltage developed by an antenna when it is aimed toward the most preferential direction. Another way of describing the half power points is to refer to them as the 3-dB points, because half power corresponds to -3 dB on the dB scale (see Figs. 7-4 and 7-5).

ANTENNA RESISTANCE

If power is to get to an antenna, it must be connected to a transmission line. It is necessary that the characteristic impedance of the transmission line be equal to the resistance presented by the antenna in order to prevent standing waves from being present on the line.

If measuring the antenna resistance is attempted by putting an ohmmeter across the antenna terminals, a reading would be obtained which would indicate an open circuit because an ohmmeter uses a dc source in measuring resistance. Therefore something other than that measured by an ohmmeter must be involved. The resistance presented by the antenna consists mainly of what is called *radiation resistance.*

The radiation resistance of an antenna is defined as a fictitious resistance which would dissipate as much power as the antenna in question is radiating if it were connected to the same transmission line (see Fig. 7-6). An antenna which is radiating 100 W when drawing 2 A has a radiation resistance of $100/2^2$, or 25 Ω $(P/I^2 = R)$. Do not lose sight of the fact that this is not a true resistance. A true resistance causes heat losses. There *are* heat losses involved in an antenna. The heat losses are not what is accounted for by the radiation resistance. The radiation resistance accounts for the power which is radiated. There is another resistance associated with an antenna which is defined to take this heat loss into account. This is called the *ohmic resistance* of the antenna. It represents the actual losses caused by the conversion of electrical energy to heat as a result of the resistivity of the various conducting elements of the antenna.

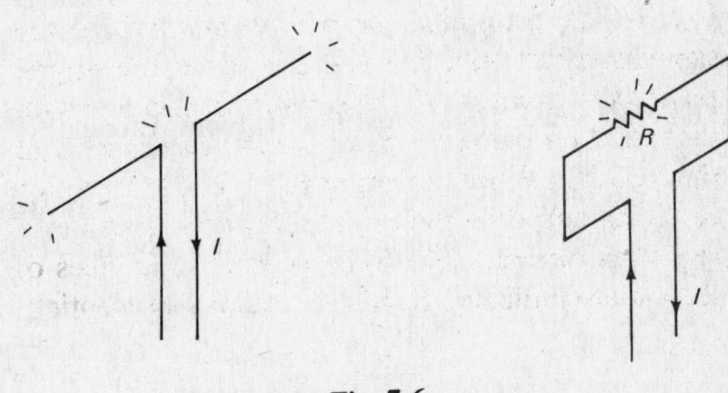

Fig. 7-6

THE ANTENNA AS A RESONANT CIRCUIT

The impedance presented by an antenna also has a reactive component due to currents and voltages being out of phase. The reason that currents and voltages are out of phase can be due to the antenna not being cut to the exact length prescribed for the type of antenna being considered. Actually an antenna is very much like a tuned circuit in that at its center frequency, that frequency at which its geometry is exactly correct, a maximum impedance which is purely resistive (radiation resistance plus ohmic resistance) is presented to the transmission line. As the frequency of the signal being presented to the antenna is changed, the impedance presented to the transmission line becomes reactive, just as with a tuned circuit. The bandwidth and Q of an antenna can be discussed just as is done with tuned circuits, and these terms still maintain their original meaning. The relationship between the bandwidth and the Q of an antenna is the same as that for tuned circuits.

$$\text{Bandwidth} = \frac{f_0}{Q}$$

If for any reason the resonant frequency of an antenna is not exactly at the frequency to be transmitted, the antenna can be tuned. *Antenna tuners* are used for this purpose. Antenna tuners are tunable reactive circuits which are tuned so that when combined with the antenna, the reactive components of the antenna and the tuning circuit cancel out. Capacitive reactance can counterbalance inductive reactance, and inductive reactance can counterbalance capacitive reactance.

VELOCITY FACTOR

As seen in Chapter 6, electromagnetic waves do not travel at the same speed in all media. The figures 186 000 miles/s or 3×10^8 meters/s for the speed of electromagnetic radiation are valid only for free space.

The velocity of electromagnetic waves is slightly different in a conductor such as the material the antenna is made of: aluminum, copper, etc. As mentioned before, in order to take this into account, the velocity factor is used. The velocity factor is that number which when multiplied by the speed of light in free space gives us the speed of light in the medium in question.

ANTENNA TYPES

The Half-Wave Dipole and the Marconi Antenna

Two of the more basic antennas are the half-wave dipole or Hertz antenna and the quarter-wave vertical or Marconi antenna. The half-wave dipole is shown as Fig. 7-7, while the quarter-wave vertical is shown as Fig. 7-8.

Fig. 7-7

Fig. 7-8

As is obvious from their names, the optimum length for each of the two antennas is one half wavelength for the Hertz and one quarter wavelength for the Marconi antenna.

Beam Antennas

A beam antenna is an antenna which has highly directional properties, essentially radiating a beam of electromagnetic radiation.

The *Yagi-Uda antenna* is frequently referred to as a beam antenna because of its highly directional radiation pattern.

There are many other antennas which have highly directional characteristics and are considered beam antennas. One of these is the *rhombic antenna*, which is shown in Fig. 7-9. It is constructed in a horizontal plane and has a radiation pattern as shown. The input resistance of the rhombic antenna is 800 Ω.

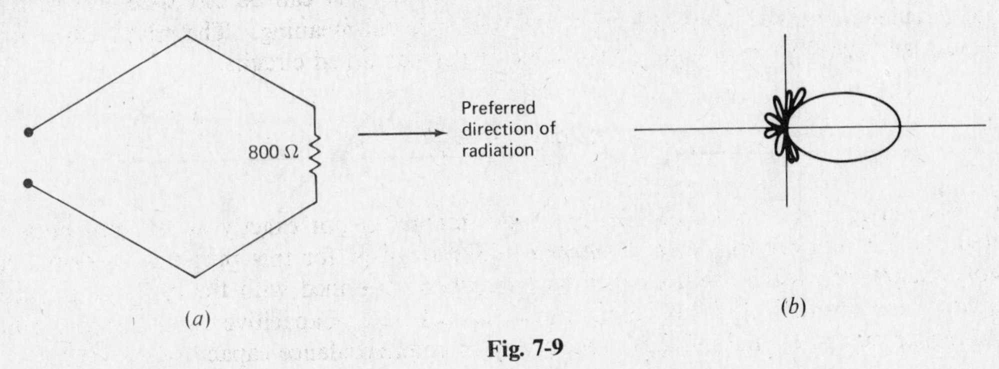

Fig. 7-9

Folded Dipole Antenna

While we are discussing other antennas, let us consider the folded dipole, shown in Fig. 7-10. The folded dipole is really a variation of the dipole antenna; the folded dipole has an input impedance of approximately 280 Ω and can be used quite conveniently with flat ribbon-type transmission line (twinex), which has a characteristic impedance of 300 Ω. The folded dipole with a reflector and twinex transmission line is quite popular for use with home-type television receivers.

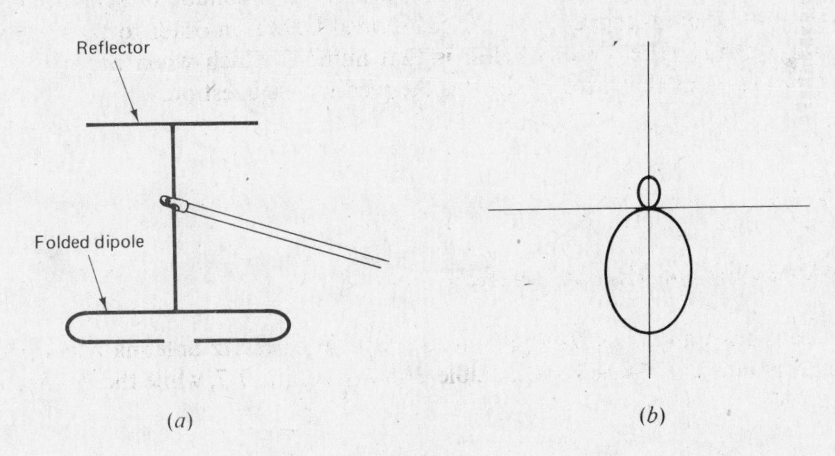

Fig. 7-10

Turnstile Antenna

An interesting variation on the dipole antenna is called the turnstile antenna. Recall the figure-eight, double-lobe pattern of the dipole antenna. Consider the radiation pattern that would result if two dipoles were constructed on the same mast perpendicular to each other, one radiating preferentially in the north-south direction and the other radiating preferentially in the east-west direction. A double figure-eight would present itself, and where the two overlapped we would have to do vector addition of the two fields; the result would be an almost circular radiation in the horizontal plane (see Fig. 7-11). The radiation patterns indicated are based upon the two dipoles being fed by signals 90° out of phase with each other but identical in all other respects. The turnstile antenna has an input resistance of 36 Ω (1/2 of that of a dipole, 72 Ω).

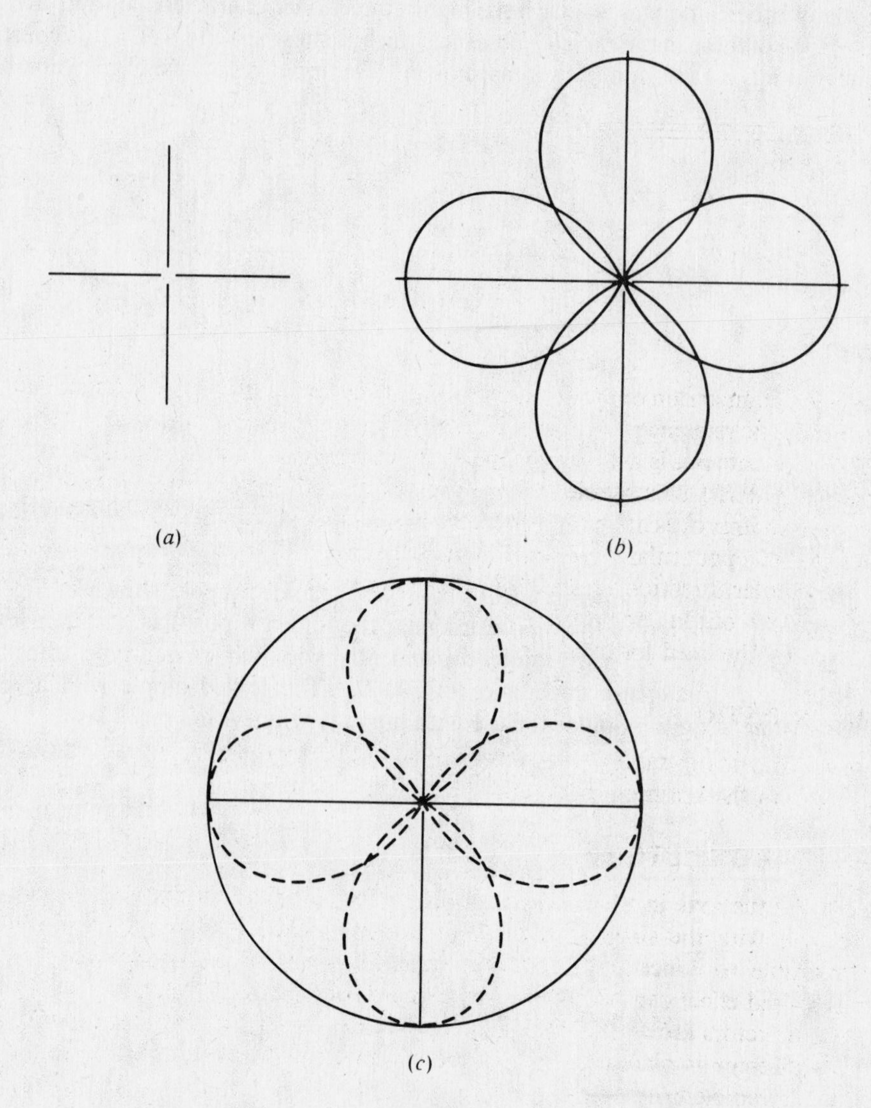

Fig. 7-11

Loop Antenna

An antenna which is used very frequently but almost entirely as a reception antenna is the loop antenna. This is the antenna which is frequently found on the back of ac-dc table radios. It can be a square loop or an oblong loop. The number of turns of wire can be anywhere from one to dozens in a coil pasted to the inside of the cabinet or to the fiberboard back. This antenna and its radiation pattern are shown in Fig. 7-12.

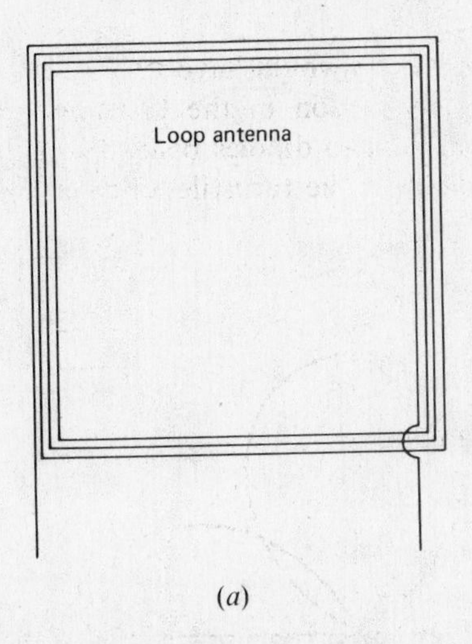

(a)

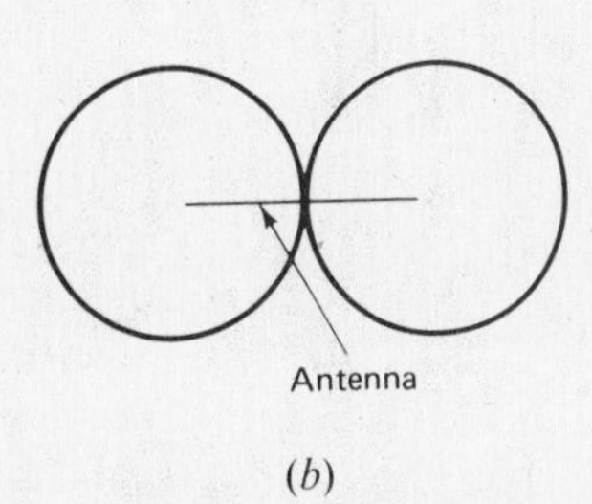

(b)

Fig. 7-12

ANTENNA GAIN

Antenna gain is a comparison of the output, in a particular direction, of the antenna in question and a *reference antenna.* The reference antenna is generally either an omnidirectional antenna, of which the Marconi quarter-wave antenna is an example, putting out equal amounts of radiation in all directions (a circular radiation pattern), or a dipole. Therefore, if it is said that an antenna has a gain of 10 dB, the antenna in question improves upon the reference antenna in that direction by 10 dB. The increased power being radiated in a particular direction is obtained at the expense of the other directions. Power is radiated in a particular direction by stealing it from other directions. Thus, antenna gain does *not* refer to obtaining more output power than input power. Frequently one antenna is compared to another, thus avoiding the need for a reference.

FRONT-TO-BACK RATIO

Front-to-back ratio is the ratio expressed in dB of output in the most optimum direction to the output 180° away from the optimum direction.

REFLECTORS AND DIRECTORS

There are many situations in which it is desirable to focus the radiated power into a more limited area than is possible with the simple dipole. For example, in radio communications between two stations, it is desirable to concentrate the total radiated power of the transmitting station in one direction. This desired effect can be obtained by using reflectors and directors.

Reflectors and directors are additional conducting elements used to obtain improved directivity of an antenna. The director is placed in front of the driven element (the dipole), while the reflector is placed behind the driven element.

Figure 7-13 is a drawing of an antenna which consists of one dipole, one reflector, and one director. Note that the director is less than one half-wavelength long and is placed closer to the dipole than is the reflector. The optimum spacing between the dipole and the director is 0.1 wavelength when the length of the director is 5% smaller than that of the dipole.

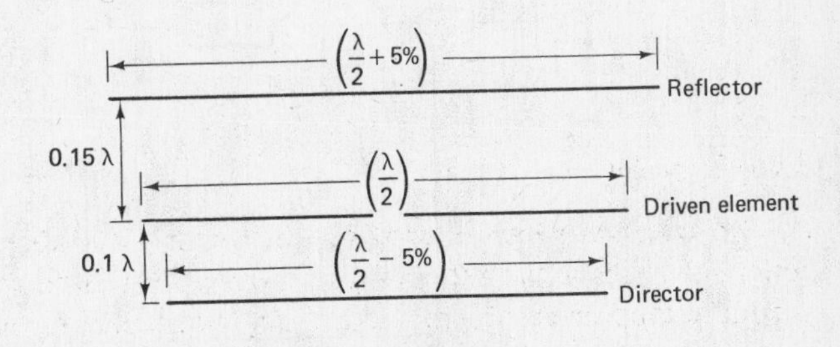

Fig. 7-13

The reflector is slightly longer than one half-wavelength and is placed less than one

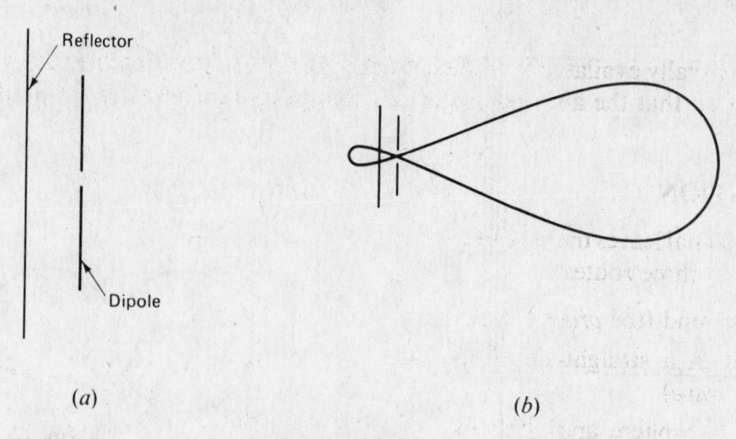

(a) (b)

Fig. 7-14

quarter-wavelength behind and parallel to the dipole. The optimum spacing is 0.15 wavelength if the reflector is 5% longer than the dipole. Figure 7-14 shows the radiation pattern of the dipole and reflector combination. Note the increased radiation in one direction at the expense of the other directions.

Directors and reflectors are known as *parasitic elements* because when used in a transmitting antenna they are not connected to the transmission line, but get their energy from the power transmitted by the dipole, which is connected to the transmission line. The dipole is known as the *driven element*. An antenna which consists of a driven element and any number of parasitic elements is known as a *parasitic array*. A dipole with reflectors and directors is frequently called a Yagi-Uda antenna, or simply a Yagi antenna.

ANTENNA TRAPS

There are situations in which one station may transmit at different frequencies at different times, making it desirable to be able to radiate efficiently at each of these frequencies. One possibility is to make use of the same transmission line and antenna but lengthen the antenna when signals of longer wavelength are to be transmitted. This can be done by placing LC circuits at various points on the antenna. These LC circuits have an impedance which ranges from very low to very high values. By choosing the L's and C's and the Q of the combination properly, a unit called a trap can be designed which behaves as an open circuit over a range of frequencies, thereby removing the portion of the antenna between it and the end, the only section of the antenna which is active being that portion between the driving point and the traps. In a center-fed dipole, traps are encountered in pairs, one in each of the quarter-wavelength sections, and the antenna length is taken as the distance between the two traps when the antenna is operating at frequencies at which the traps are considered open circuits (see Fig. 7-15).

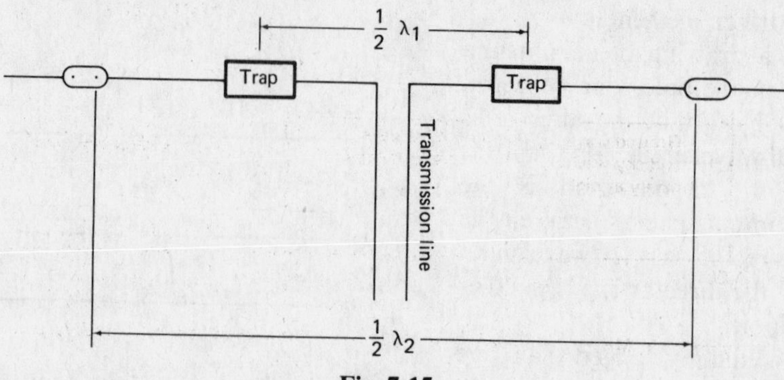

Fig. 7-15

Traps are commercially available for different frequencies. It is acceptable to have more than one set of traps on a line so that the antenna length can change to take on many different values.

WAVE PROPAGATION

Once a radiated signal leaves the antenna, it travels along one of three routes:

1. Along the ground (the *ground wave*)
2. Straight out in a straight line (the *line-of-sight wave*)
3. Up to the ionosphere and back to earth (the *sky wave*)

See Fig. 7-16.

The frequency of the signal determines which of these modes predominates. See Fig. 7-17.

The effectiveness of the line-of-sight wave is, as its name implies, limited to a line of sight between the transmitting and receiving antennas.

An equation which can be used to calculate the maximum distance between transmitting and receiving antennas for direct line-of-sight waves to be effective is

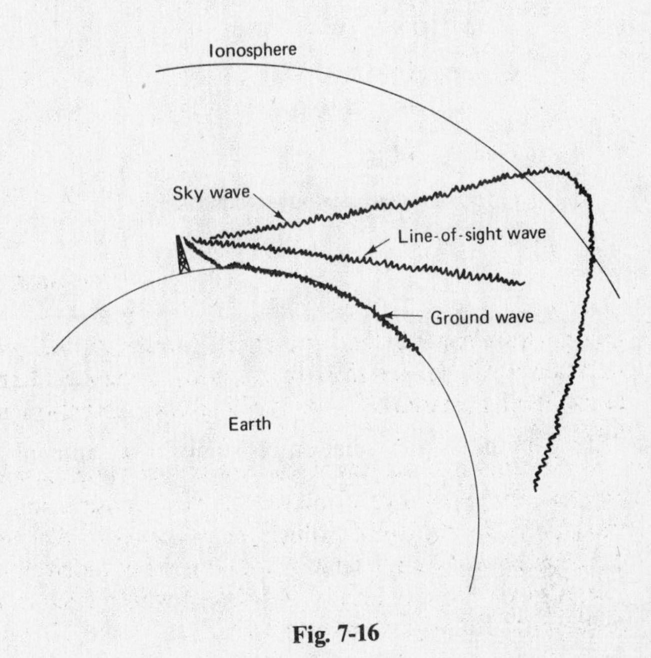

Fig. 7-16

$$d = \sqrt{2h_t} + \sqrt{2h_r}$$

where h_t is the height of the transmitting antenna in feet, h_r is the height of the receiving antenna in feet, and d is the maximum distance in miles over which communication between them can take place by direct line-of-sight wave.

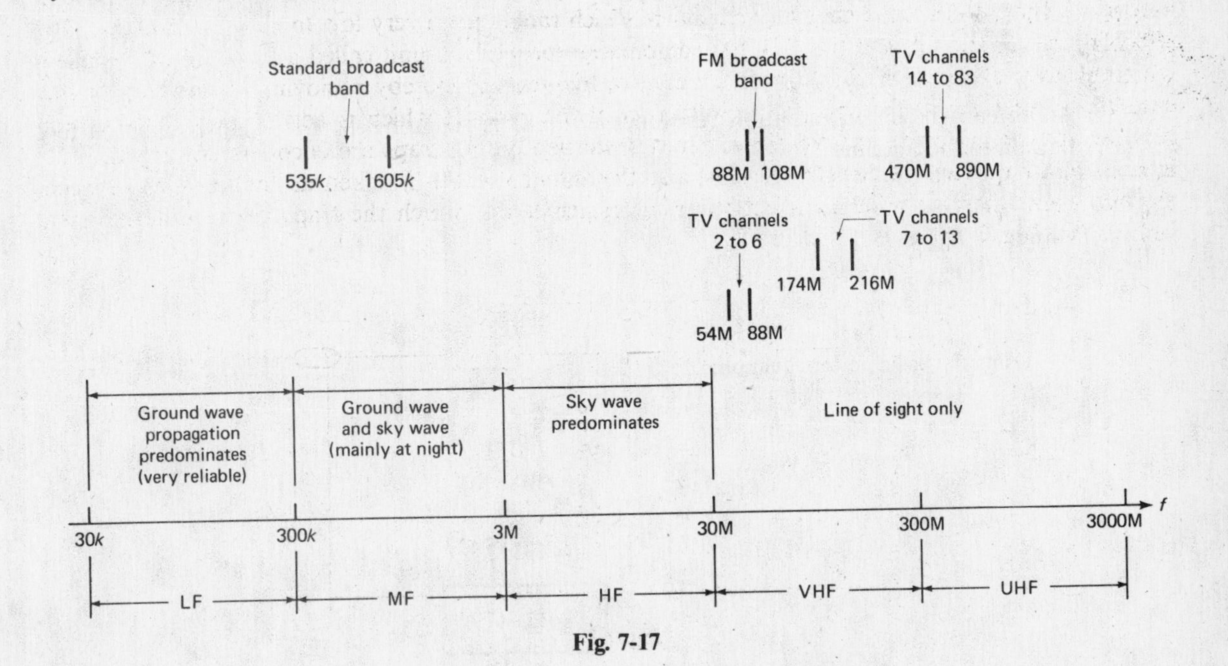

Fig. 7-17

Solved Problems

7.1 How much power will an antenna having a radiation resistance of 50 Ω radiate when it is fed 20 A?

SOLUTION

Given: $I = 20$ A

$R_{rad} = 50\ \Omega$

Find: P_{rad}

Using the basic power equation,

$$P = I^2R$$
$$= (20)^2(50)$$
$$\boxed{P = 20\,000 \text{ W}}$$

7.2 What is the radiation resistance of an antenna which radiates 5 kW when it draws 15 A?

SOLUTION

Given: $P = 5$ kW

$I = 15$ A

Find: R_{rad}

From the basic power equation,

$$P = I^2R$$
$$5000 = (15)^2R_{rad}$$
$$R_{rad} = \frac{5000}{(15)^2}$$
$$\boxed{R_{rad} = 22.2\ \Omega}$$

7.3 An antenna having a radiation resistance of 75 Ω is radiating 10 kW. How much current flows into the antenna?

SOLUTION

Given: $R_{rad} = 75\ \Omega$

$P_{rad} = 10$ kW

Find: I

Using the basic power equation,

$$P = I^2R$$
$$10 \times 10^3 = I^2(75)$$
$$I^2 = \frac{10 \times 10^3}{75}$$
$$= 133.33$$
$$\boxed{I = 11.547 \text{ A}}$$

7.4 Determine the Q of an antenna if it has a bandwidth of 0.6 MHz and is cut to a frequency of 30 MHz.

SOLUTION

Given: $BW = 0.6$ MHz

$f_0 = 30$ MHz

Find: Q

Using the BW equation,

$$BW = \frac{f_0}{Q}$$

$$0.6 \times 10^6 = \frac{30 \times 10^6}{Q}$$

$$Q = \frac{30 \times 10^6}{0.6 \times 10^6}$$

$$\boxed{Q = 50}$$

7.5 Determine the length of a half-wave dipole antenna to be used to receive a 5-MHz radio signal. Assume that the velocity of electromagnetic waves on the antenna is 3×10^8 meters/s. See Fig. 7-18.

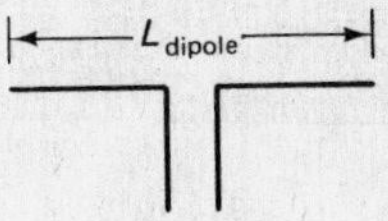

SOLUTION

L_{dipole}

Fig. 7-18

Given: $f = 5$ MHz

Dipole antenna

$C = 3 \times 10^8$ m/s

Find: L_{dipole}

The length of a dipole antenna is equal to one half-wavelength. Finding the wavelength of the signal on the antenna is the first order of business:

$$f\lambda = C$$

$$5 \times 10^6 \lambda = 3 \times 10^8$$

$$\lambda = \frac{3 \times 10^8}{5 \times 10^6}$$

$$= 60 \text{ m}$$

The wavelength must now be divided by 2 in order to determine the required length of the dipole.

$$L_{dipole} = \frac{\lambda}{2} = \frac{60}{2}$$

$$\boxed{L_{dipole} = 30 \text{ m}}$$

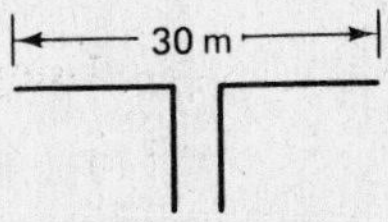

30 m

Fig. 7-19

See Fig. 7-19.

7.6 A half-wave dipole antenna is to be cut to optimally radiate a 250-MHz signal. The velocity factor of the antenna elements is 0.85. Determine the necessary length of the antenna when taking the velocity factor into account, and then determine what the length would be if the velocity factor were 1.0.

SOLUTION

Given: $f = 250$ MHz

$k_1 = 0.85$

$k_2 = 1.0$

Find: L_1, L_2

$$f\lambda_1 = k_1 C$$

$$250 \times 10^6 \lambda_1 = 0.85(3 \times 10^8)$$

$$\lambda_1 = \frac{0.85(3 \times 10^8)}{250 \times 10^6}$$

$$= 1.02 \text{ m}$$

$$L_1 = \frac{\lambda_1}{2}$$

$$= \frac{1.02}{2}$$

$$\boxed{L_1 = 0.51 \text{ m}}$$

$$f\lambda_2 = k_2 C$$

$$250 \times 10^6 \lambda_2 = (1)(3 \times 10^8)$$

$$\lambda_2 = \frac{3 \times 10^8}{250 \times 10^6}$$

$$= 1.20 \text{ m}$$

$$L_2 = \frac{\lambda_2}{2}$$

$$= \frac{1.2}{2}$$

$$\boxed{L_2 = 0.6 \text{ m}}$$

7.7 An antenna is needed for the transmission of a signal having a center frequency of 60 MHz. Determine the length of a Hertz dipole antenna suited for this purpose. Assume a velocity factor of 0.85 for the antenna conductors. See Fig. 7-20.

SOLUTION

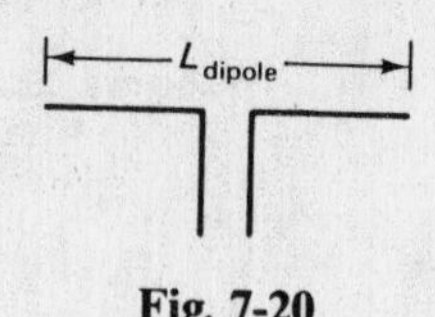

Fig. 7-20

Given: $f = 60$ MHz

$k = 0.85$

Find: L_{dipole}

The length of a Hertz dipole antenna should be equal to half the wavelength of the signal on the antenna. Thus we must find the wavelength of the signal on the antenna.

$$f\lambda = kC$$

$$60 \times 10^6 \lambda = 0.85(3 \times 10^8)$$

$$\lambda = \frac{0.85(3 \times 10^8)}{60 \times 10^6}$$

$$= 0.0425 \times 10^2$$

$$= 4.25 \text{ m}$$

The length of the Hertz dipole is $\lambda/2$. Thus,

$$\frac{\lambda}{2} = \frac{4.25}{2}$$

$$\boxed{\frac{\lambda}{2} = 2.125 \text{ m}}$$

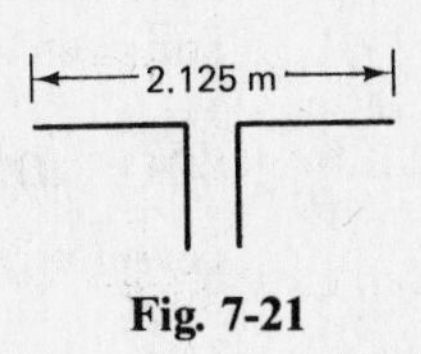

Fig. 7-21

See Fig. 7-21.

7.8 Determine the optimum length of a Marconi antenna for the transmission of a 100-MHz signal. The velocity factor for the antenna is 0.90. See Fig. 7-22.

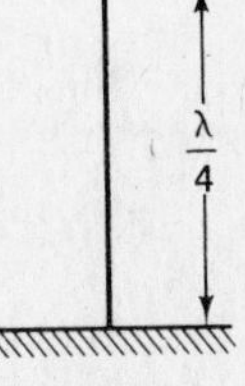

Fig. 7-22

SOLUTION

Given: $f = 100$ MHz

$k = 0.90$

Find: L_{Marconi}

The optimum length of a Marconi vertical antenna is one quarter-wavelength. Finding the wavelength,

$$f\lambda = kC$$

$$100 \times 10^6 \lambda = 0.90(3 \times 10^8)$$

$$\lambda = \frac{0.90(3 \times 10^8)}{100 \times 10^6}$$

$$= 0.027 \times 10^2$$

$$= 2.7 \text{ m}$$

One quarter wavelength is then

$$\boxed{\frac{\lambda}{4} = \frac{2.7}{4} = 0.675 \text{ m}}$$

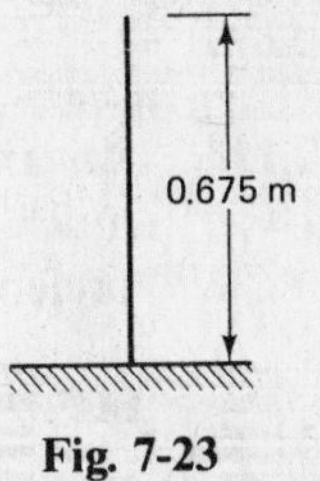

Fig. 7-23

See Fig. 7-23.

7.9 Find the intended frequency of operation of a dipole antenna cut to a length of 3.5 m. Assume a velocity of 3×10^8 m/s for electromagnetic waves on the antenna.

SOLUTION

Given: $L_{\text{dipole}} = 3.5$ m

Find: f_0

A dipole antenna is cut to one half-wavelength. Thus,

$$\lambda = 2L_{\text{dipole}} = 2(3.5) = 7.0 \text{ m}$$

Converting the wavelength to frequency,

$$f\lambda = C$$

$$f(7.0) = 3 \times 10^8$$

$$f = \frac{3 \times 10^8}{7}$$

$$= 0.429 \times 10^8$$

$$\boxed{f = 42.9 \text{ MHz}}$$

7.10 Determine the dB gain of a receiving antenna which delivers a 40-μV signal to a transmission line over that of an antenna that delivers a 20-μV signal under identical circumstances.

SOLUTION

Given: $V_2 = 40\ \mu V$
$V_1 = 20\ \mu V$

Find: A_{dB}

Using the dB-gain formula,

$$A_{dB} = 20 \log_{10} \frac{V_2}{V_1}$$

Substituting,

$$A_{dB} = 20 \log_{10} \frac{40 \times 10^{-6}}{20 \times 10^{-6}}$$
$$= 20 \log_{10} 2$$

Referring to a table of logarithms or using an electronic calculator, $\log_{10} 2 = 0.3$. Thus,

$$A_{dB} = 20(0.3)$$

$$\boxed{A_{dB} = 6\ \text{dB}}$$

7.11 An antenna that has a gain of 6 dB over a reference antenna is radiating 700 W. How much power must the reference antenna radiate in order to be equally effective in the most preferred direction?

SOLUTION

Given: $A_{dB} = 6\ \text{dB}$
$P_1 = 700\ \text{W}$

Find: P_2

The reference antenna must supply an output 6 dB higher than 700 W. Using the dB-gain formula,

$$A_{dB_{10}} = 10 \log_{10} \frac{P_2}{P_1}$$
$$6 = 10 \log_{10} \frac{P_2}{700}$$
$$0.6 = \log_{10} \frac{P_2}{700}$$
$$\text{antilog } 0.6 = \frac{P_2}{700}$$
$$4 = \frac{P_2}{700}$$
$$P_2 = 4(700)$$

$$\boxed{P_2 = 2800\ \text{W}}$$

7.12 Determine the front-to-back ratio of an antenna which puts out 3 kW in its most optimum direction and 500 W in the opposite direction. See Fig. 7-24.

SOLUTION

Given: $P_F = 3$ kW

$P_B = 500$ W

Find: A_{FB}

The front-to-back ratio is the ratio of powers of the optimum direction to its opposite direction expressed in dB.

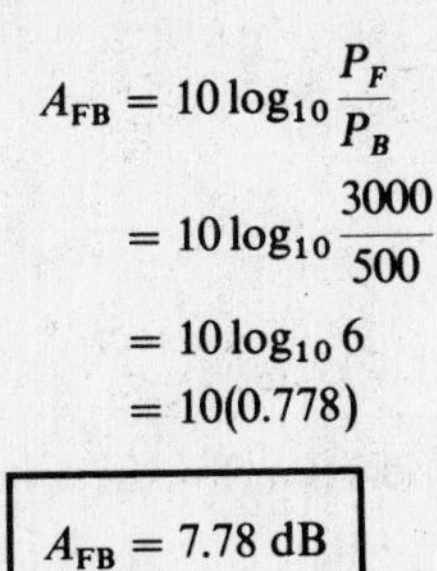

$$A_{FB} = 10 \log_{10} \frac{P_F}{P_B}$$

$$= 10 \log_{10} \frac{3000}{500}$$

$$= 10 \log_{10} 6$$

$$= 10(0.778)$$

$$\boxed{A_{FB} = 7.78 \text{ dB}}$$

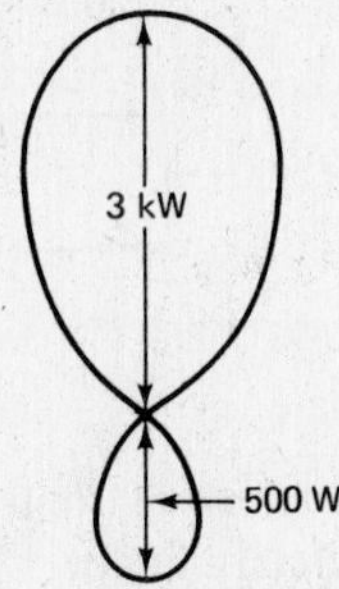

Fig. 7-24

7.13 Design a beam antenna consisting of a dipole, one reflector, and one director. Assume a velocity coefficient of 0.90. The antenna is to be cut for a frequency of 150 MHz. See Fig. 7-25.

SOLUTION

Given: $f = 150$ MHz

$k = 0.90$

Find: Beam antenna specifications

It is first necessary to determine the wavelength of the 150-MHz signal on the antenna.

$$f\lambda = kC$$

$$(150 \times 10^6)\lambda = 0.90(3 \times 10^8)$$

$$\lambda = \frac{0.90(3 \times 10^8)}{150 \times 10^6}$$

$$= 1.8 \text{ m}$$

$(1 - .05)\left(\frac{\lambda}{2}\right)$

$0.15\,\lambda$

$\left(\frac{\lambda}{2}\right)$

$(1 + .05)\left(\frac{\lambda}{2}\right)$

$0.1\,\lambda$

Fig. 7-25

Now, to determine the length of the antenna elements:

$$L_{\text{dipole}} = \frac{\lambda}{2} = \frac{1.8}{2}$$

$$\boxed{L_{\text{dipole}} = 0.9 \text{ m}}$$

$$L_{\text{reflector}} = (1 + 0.05)\frac{\lambda}{2}$$

$$= (1.05)(0.9)$$

$$\boxed{L_{\text{reflector}} = 0.945 \text{ m}}$$

$$L_{\text{director}} = (1 - 0.05)\frac{\lambda}{2}$$
$$= (0.95)(0.9)$$

$$\boxed{L_{\text{director}} = 0.855 \text{ m}}$$

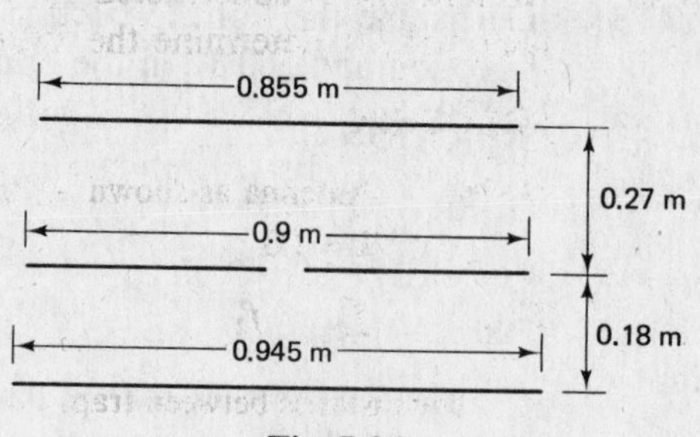

Fig. 7-26

Now determine element spacing:

$$\boxed{0.1\lambda = (0.1)(1.8) = 0.18 \text{ m}}$$

$$\boxed{0.15\lambda = (0.15)(1.8) = 0.27 \text{ m}}$$

See Fig. 7-26.

7.14 Determine the optimum frequencies of operation for the antenna shown as Fig. 7-27. Assume a velocity factor of 1.0.

SOLUTION

Given: Antenna shown in Fig. 7-27

Find: f_1, f_2

The antenna shown in Fig. 7-27 is a quarter-wave Marconi antenna. The antenna is cut for two frequencies, one whose wavelength is four times the length between the feed point and the antenna trap, and the other four times the total length of the antenna.

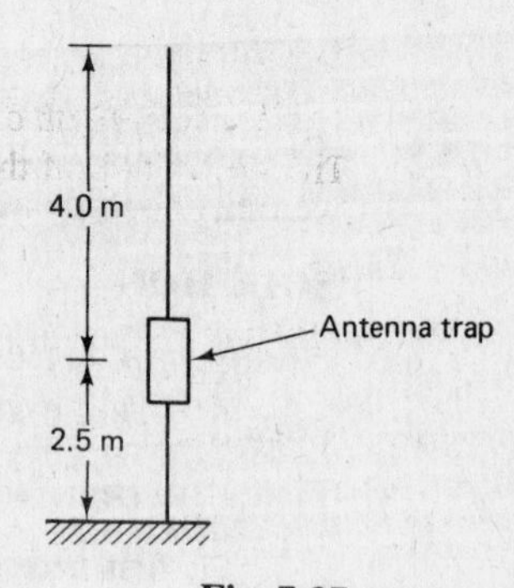

Fig. 7-27

$$\lambda_1 = 4 \times 2.5 = 10 \text{ m}$$
$$= 10 \text{ m}$$

$$\lambda_2 = 4(2.5 + 4.0)$$
$$= 26 \text{ m}$$

Converting these wavelengths to frequencies requires the use of the formula

$$f\lambda = kC$$

Thus,

$$f_1\lambda_1 = kC$$
$$f_1(10) = (1)(3 \times 10^8)$$

$$\boxed{f_1 = 30 \text{ MHz}}$$

$$f_2\lambda_2 = kC$$
$$f_2(26) = (1)(3 \times 10^8)$$
$$f_2 = \frac{3 \times 10^8}{26}$$

$$\boxed{f_2 = 11.54 \text{ MHz}}$$

7.15 An antenna is constructed as shown in Fig. 7-28 using antenna traps. Assuming a velocity factor of 1.0, determine the frequencies at which this antenna has been designed to operate.

SOLUTION

Given: Antenna as shown in Fig. 7-28
$k = 1.0$

Find: f_1, f_2

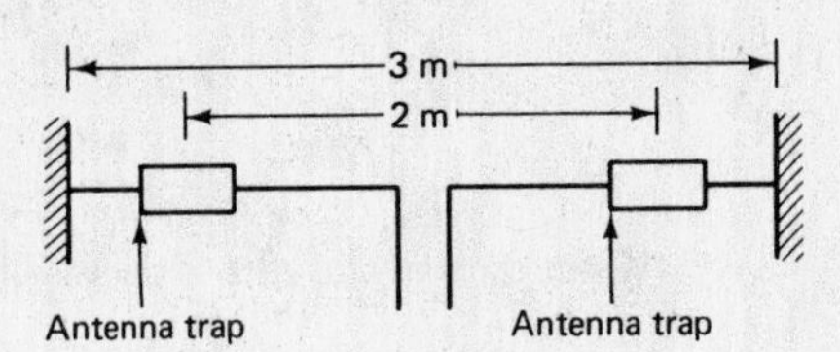

Fig. 7-28

The distance between traps is equal to 1/2 wavelength of the higher optimum frequency of operation. Thus the wavelength of the higher-frequency signal is equal to $2 \times 2\,\text{m} = 4\,\text{m}$.

Converting wavelength to frequency,

$$f_1\lambda_1 = kC$$
$$f_1(4) = 1(3 \times 10^8)$$
$$f_1 = \frac{3 \times 10^8}{4}$$
$$= 0.75 \times 10^8$$

$$\boxed{f_1 = 75\ \text{MHz}}$$

The full-length distance of the antenna is 1/2 wavelength of the lower optimum frequency of operation. The wavelength of the lower-frequency signal is thus equal to 6 m.

$$f_2\lambda_2 = kC$$
$$f_2(6) = 1(3 \times 10^8)$$
$$f_2 = \frac{3 \times 10^8}{6}$$
$$= 0.5 \times 10^8$$

$$\boxed{f_2 = 50\ \text{MHz}}$$

7.16 A dipole antenna using antenna traps is desired which is to be used to transmit a 140-MHz signal and a 90-MHz signal. See Fig. 7-29. Design the antenna. Assume a velocity factor of 0.90.

SOLUTION

Given: $f_1 = 140$ MHz
$f_2 = 90$ MHz
$k = 0.90$

Find: L_1, L_2

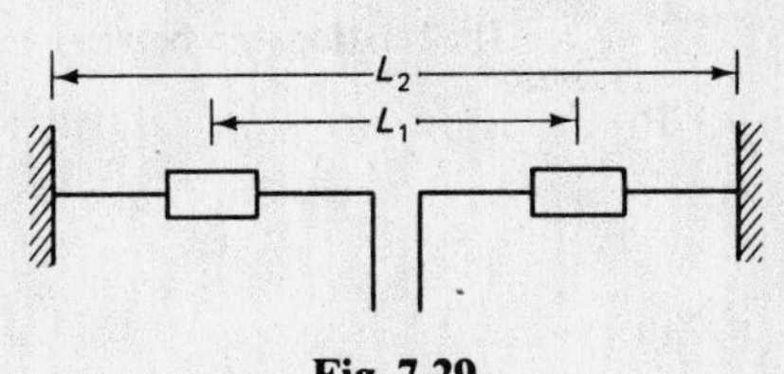

Fig. 7-29

First find the wavelengths of the two signals to be radiated:

$$f_1\lambda_1 = kC$$
$$140 \times 10^6\lambda_1 = 0.90(3 \times 10^8)$$
$$\lambda_1 = \frac{0.90(3 \times 10^8)}{140 \times 10^6}$$
$$= 1.929\ \text{m}$$

$$f_2 \lambda_2 = kC$$
$$90 \times 10^6 \lambda_2 = 0.90(3 \times 10^8)$$
$$\lambda_2 = \frac{0.90(3 \times 10^8)}{90 \times 10^6}$$
$$= 3.0 \text{ m}$$

Once the wavelengths of the two signals have been found, the required antenna lengths can be determined:

$$L_1 = \frac{\lambda_1}{2}$$
$$= \frac{1.929}{2}$$

$$\boxed{L_1 = 0.965 \text{ m}}$$

$$L_2 = \frac{\lambda_2}{2}$$
$$= \frac{3.0}{2}$$

$$\boxed{L_2 = 1.5 \text{ m}}$$

Fig. 7-30

See Fig. 7-30.

7.17 An antenna is to be installed to receive a line-of-sight wave transmitted from an antenna located at a distance of 50 mi from this installation and which is 750 ft in height. Determine the minimum necessary height of the receiving antenna.

SOLUTION

Given: $d = 50$ mi

$h_t = 750$ ft

Find: h_r

The relationship between antenna height in feet and line-of-sight distance in miles is

$$d = \sqrt{2h_t} + \sqrt{2h_r}$$
$$50 = \sqrt{2(750)} + \sqrt{2h_r}$$
$$50 - 38.73 = \sqrt{2h_r}$$
$$11.27 = \sqrt{2h_r}$$
$$127 = 2h_r$$
$$\frac{127}{2} = h_r$$
$$63.5 = h_r$$

$$\boxed{h_r = 63.5 \text{ ft}}$$

The receiving antenna must be higher than 63.5 ft.

7.18 How far from a transmitting antenna 1500 ft high can a line-of-sight wave be effective? Assume a receiving antenna height of 25 ft. See Fig. 7-31.

SOLUTION

Given: $h_t = 1500$ ft

$h_r = 25$ ft

Find: d

Using the formula relating distance in miles and antenna height in feet,

$$\begin{aligned} d &= \sqrt{2h_t} + \sqrt{2h_r} \\ &= \sqrt{2(1500)} + \sqrt{2(25)} \\ &= \sqrt{3000} + \sqrt{50} \\ &= 54.77 + 7.07 \end{aligned}$$

$d = 61.84$ mi, assuming no obstructions

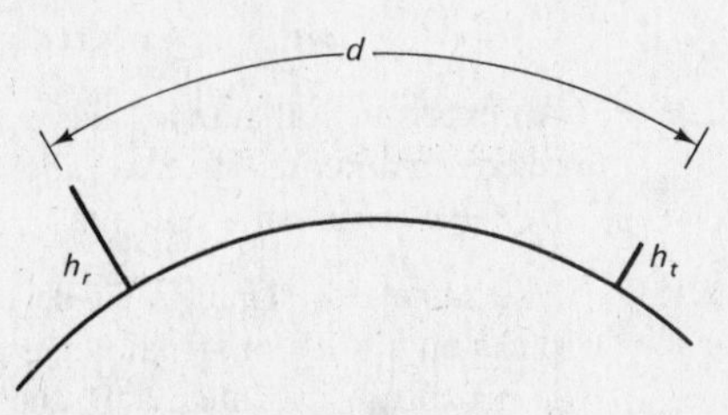

Fig. 7-31

Supplementary Problems

7.19 Sketch the radiation pattern of a Marconi antenna. *Ans.* See Fig. 7-1.

7.20 Sketch the radiation pattern of a Hertz antenna. *Ans.* See Fig. 7-2.

7.21 What is another name for a Hertz antenna? *Ans.* Half-wave dipole.

7.22 Distinguish between the radiation resistance of an antenna and the ohmic resistance of an antenna.
Ans. Radiation resistance accounts for radiated power. Ohmic resistance accounts for heat generated due to resistivity.

7.23 Determine the radiation resistance of an antenna which radiates 1000 W when drawing 5 A.
Ans. 40 Ω

7.24 How much current does an antenna draw when radiating 500 W if it has a radiation resistance of 300 Ω?
Ans. 1.29 A

7.25 An antenna having a radiation resistance of 50 Ω draws 8 A. How much power is it radiating?
Ans. 3200 W

7.26 Calculate the radiation resistance of an antenna which is fed 12 A when radiating 10 kW.
Ans. 69.44 Ω

7.27 How much power does a 50-Ω antenna radiate when fed a current of 4 A? *Ans.* 800 W

7.28 Determine the Q of an antenna cut for a frequency of 14 MHz having a bandwidth of 4 MHz.
Ans. $Q = 3.5$

7.29 What is the bandwidth of an antenna cut for 110 MHz having a Q of 70?
Ans. BW = 1.57 MHz

7.30 An antenna cut for a frequency of 30 MHz has a Q of 40. Determine its bandwidth.
Ans. BW = 750 kHz

7.31 Calculate the length of a half-wave dipole antenna cut for a frequency of 75 MHz. Assume a velocity factor of 0.90. *Ans.* 1.8 m

7.32 Assuming a velocity factor of 0.88, determine the optimum length of a Marconi antenna which is to be cut for a frequency of 65 MHz. *Ans.* 1.015 m

7.33 Calculate the required length of a half-wave dipole antenna cut for 90 MHz if the velocity factor on the antenna is 0.87. Then determine the required length of the antenna if the velocity factor were 1.0.
Ans. 1.45 m, 1.667 m

7.34 A Marconi antenna is 2.8 m in length. The velocity factor is 0.80. What frequency was this antenna cut for? *Ans.* 21.43 MHz

7.35 An experimental antenna has a gain of 3 dB above a reference antenna. How much power would the reference antenna have to radiate in order to provide the same signal picked up when 5 W is radiated by the experimental antenna? *Ans.* 10 W

7.36 The same test signal is to be radiated by antenna A and then by antenna B. A receiving antenna picks up a 5-μV/m signal when the transmission is broadcast by antenna A. When the same test signal is broadcast by antenna B, the receiving antenna picks up a 20-μV/m signal. Calculate the gain of antenna B over antenna A. *Ans.* 12 dB

7.37 An antenna having a gain of 3 dB over a reference antenna is radiating 1500 W. How much power must the reference antenna radiate in order to be equally as effective in the most preferred direction?
Ans. 3000 W

7.38 Calculate the front-to-back ratio of an antenna which radiates 500 W in a northerly direction and 50 W in a southerly direction. *Ans.* 10 dB

7.39 Sketch an antenna showing one driven element, one reflector, and one director. Assume the speed of propagation of the electromagnetic wave on the antenna to be 3×10^8 m/s. Determine the length of each of the antenna elements and the distance between them if this antenna is optimized for 40 MHz.
Ans. 3.75 m, 3.9375 m, 3.5625 m, 0.75 m, 1.125 m

7.40 Design a three-element beam antenna (one driven element, two parasitic elements) for use with a 200-MHz signal. Consider that the velocity factor on the antenna is 0.85.
Ans. 0.6375 m, 0.669 m, 0.606 m, 0.1275 m, 0.1913 m

7.41 The driven element of a three-element Yagi antenna is to be 3.5 m in length. The velocity factor for this antenna is 0.89. Determine the frequency for which this driven element is cut and calculate the length and spacing of the other antenna elements.
Ans. 38.14 MHz, 3.675 m, 3.325 m, 0.7 m, 1.05 m

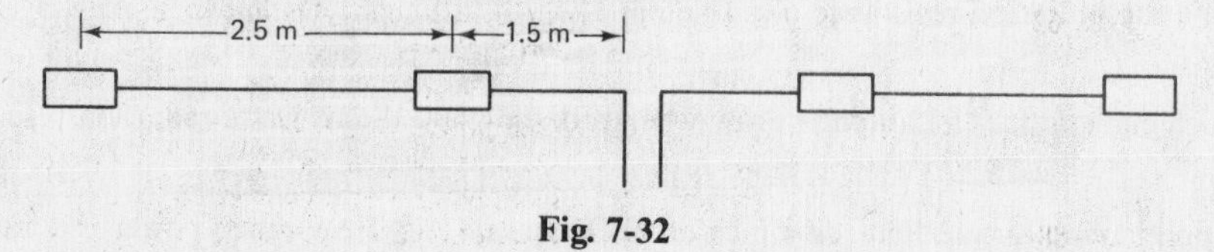

Fig. 7-32

7.42 An antenna is constructed as shown in Fig. 7-32. At what frequencies is this antenna designed to operate? Assume a velocity factor of 1.0. *Ans.* 50 MHz, 18.75 MHz

7.43 Design a dipole using traps for transmitting signals at 175 MHz and 100 MHz. and trap locations. Assume a velocity factor of 1.0. *Ans.* 0.429 m, 0.321 m
Show segment lengths

7.44 Determine the optimum frequencies of operation for the Marconi antenna shown in Fig. 7-33. Assume a velocity factor of 1.0.
Ans. 25 MHz, 9.375 MHz

7.45 Determine the maximum effective distance of the line-of-sight wave of a transmission being radiated from an antenna whose height is 1200 ft if the receiving antenna height is 100 ft. *Ans.* 63.13 mi

7.46 Calculate the necessary height of an antenna to receive a line-of-sight wave being transmitted from an antenna 1000 ft in height which is located 100 mi from the receiving antenna.
Ans. 1527.9 ft

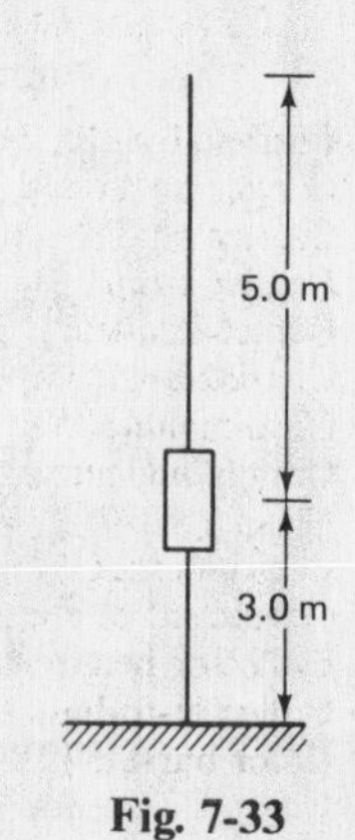

Fig. 7-33

Index